Chemistry For Students

The Only Chemistry Guide You'll Need to Ace Your Course

Leonel Travers

Table of Contents

Chapter 1: High School Chemistry Review

No one misses high school. Piles of homework, alarm bells, teachers who were as bored as you were. I'm sure you're glad to be out. But believe it or not, many things during those tedious years were invaluable. It's a shame that we have so few passionate teachers to bring such knowledge to life.

Thanks to the schooling system and a lack of enthusiasm on the part of teachers, people make the mistake of viewing chemistry as something boring, unnecessary, and even laborious. Fear not, for chemistry is only such things in the mind of the unknowing.

Once you get the axioms down and have an appreciation for all that chemistry has enabled us to do, you will enjoy it far more than before. It will provide answers to questions you didn't even know you had and open doors to worlds you completely ignored.

So, without homework to worry about, and without a teacher breathing down your neck, let's delve back into the H2O and recap the basics of this fascinating science.

The Basic Concepts of Chemistry

When we think about chemistry, the average person will probably think of white lab coats and beakers on a table. Some will probably think of the popular TV show, *Breaking Bad*.

What many do not realize is how important chemistry is as a science and how much it has helped us collectively as the human race to accomplish. Chemistry, alongside physics and mathematics, is one of the most important sciences in terms of understanding the universe and ourselves.

We often take the periodic table for granted. But it is the result of thousands of years of human discoveries placed into one colorful little chart. It's a great deal more than a chart, though. The periodic table is, in a technical sense, an array of legos from which the universe is built. The laws of physics and mathematics simply put those legos in place. Those who conceptualized and created the periodic table were true geniuses who dedicated themselves to finding out more about our reality. Of the 118 elements in the periodic table, 92 are natural. The rest are man-made, meaning that they are unstable (OpenStax, 2015).

The periodic table helps us understand and categorize what the things around us are made from. And knowing what things are made of and how they interact is vital to comprehending reality, a skill that seems to be sadly lacking in society these days.

For example, did you know that atoms are 99% space? So why don't you simply fall through your seat right now?

Atoms are comprised of subatomic particles known as neutrons, protons, and electrons. Neutrons and protons are roughly the same size at nearly one Dalton (or atomic mass unit), whereas electrons are approximately 1/1800th of their size. Because of this, only neutrons and protons really add to the mass of an atom.

Positively (+) charged protons and uncharged neutrons bond to form the nucleus of an atom, and negatively (-) charged electrons reside on the outside. The only exception to this rule is hydrogen, which contains only one proton and one electron (OpenStax, 2015). Most of the time, atoms have a balanced number of protons and electrons, meaning that their net charge is neutral. Having an equal number of positively charged particles (protons) and negatively charged particles (electrons) means that, essentially, the two cancel each other out, leaving no excess positive or negative charge. Later on, we'll look at what happens when this balance is disrupted.

Electrons, more often than not, are depicted as ring-like, orbiting the nuclei (protons and neutrons) of the atom. But newer findings have shown them to form a structure resembling that of a 'cloud' (Siegel, 2020) around the nucleus. These clouds of electrons then repel each other in a manner not unlike that of magnets to form what we perceive as solidity. However, we will be referring to the more traditional 'shell' or 'orbital' (OpenStax, 2015) depiction and understanding of electrons for the rest of this book to avoid confusion.

How do we differentiate atoms? We look at the number of protons within them. This is similar to the way in which we differentiate animals or plants from one another. Animals and plants have characteristics that differ between species and help us identify them at multiple levels of analysis.

The way we do this with elements isn't by looking at their DNA, plumage, or leaf structure. Rather, in the realm of chemistry, elements are differentiated by the number of protons in their nuclei.

Measurements and Units

Measuring quantities in any and all sciences is vital. When you take a measurement of something, a quantitive comparison is being made between the thing that you are measuring and the instrument of measurement used for that substance.

The "International System of Units" (Abozenadah et al., 2017), or SI, is the list of 7 internationally recognized units of measurement used in science. All other forms of measurement are derived from these 7.

Quantity	SI Unit (Abbreviation)
Mass	Kilogram (kg)
Length	Meter (m)
Time	Seconds (s)
Amount of substance	Mole (mol)
Temperature	Kelvin (K)
Electrical Current	Ampere (amp)
Luminous Intensity	Candella (cd)

These seven units make up what is commonly known as the metric system. In chemistry, we will typically be sticking to 5 of the 7; mole, kilogram, meter, second, and kelvin. Kelvin—the measure of absolute temperature—is often substituted with celsius (°C). These are not identical values. The difference between them can be defined as follows:

$$K = °C + 273.15$$

OR

$$°C = K - 273.15$$

When dealing with temperature, be mindful to never confuse it with heat. Temperature is the average kinetic energy of the particles that make up a material, "a measure of how hot or cold an object is relative to another object (its thermal energy content), whereas heat is the flow of thermal energy between objects with different temperatures" (LibreTexts, 2014a).

Mass in SI is measured in kilograms. Within each kilogram are 1,000 grams, and within each gram are 1,000 milligrams. In 1889, the first prototypical model of the kilogram, or IPK (International Prototype of the Kilogram), was made from platinum-iridium alloy and is currently held in Paris, France. All other kilograms are equal to this prototype (Abozenadah et al., 2017). It is important to know that mass and weight are not the same. The weight of an object is influenced by the force of gravity.

Weight= Mass x The Acceleration of Gravity

OR

W=mg (Kramer, 2017)

People often use weight and mass interchangeably in day-to-day language, but in the realm of science, it is important to understand the distinction. The mass of an object is the total amount of matter within it. It does not change regardless of where you place it. Weight, on the other hand, is the exertion of gravitational force upon that object. It *can change* depending on where you place it in the universe.

The mass of a coconut stays the same in your bedroom, at the bottom of the ocean, and on the moon. The weight of the coconut changes across those locations based on the presence of gravity. The force of gravity is not constant, not even on the earth's surface. Thus, the weight of objects is not constant and is not a reliable form of measurement.

Length is measured in meters. For those of you who are cheering for metric supremacy, it is wise to note that the metric system is just as arbitrary as the imperial system.

The meter, similar to the kilogram, has a prototypical 'model' (for lack of a better word) that is representative of all meters. Unlike the kilogram, though, it has gone through a number of changes. In 1889, it was represented by a bar of platinum-iridium alloy, then it changed to a wavelength of krypton-86 radiation in 1960, and in 1983, it changed for the final time to "the length of the path traveled by light in a vacuum during a time interval of 1/299,792,458 second" (Abozenadah et al., 2017). This information is really just background knowledge. For now, focus on the meter as the SI unit of length.

The SI unit of time is… interesting. The second (1/60 of a minute) is not the simple little tick of the clock that we normally associate with it. It has been redefined several times over the years, but we currently define it as the "duration of 9,192,631,770 periods of the radiation corresponding to the transition between the two hyperfine levels of the ground state of the cesium-133 atom" (Abozenadah et al., 2017). However, for simplicity's sake, think of it as the usual fraction of a minute.

How about measuring the amount of a substance? The mole—or mol in SI units—is used to measure very small particles in relatively large amounts. It is also known as "Avogadro's number," named after the Italian physicist Amadeo Avogadro. Why 'number' and not 'measurement' or 'unit'? Just as the word 'dozen' means twelve of a given item (for example, a dozen apples) a mole or mol refers to a set number of atoms or molecules. Unlike a dozen though, we must use scientific notation to define the number that constitutes one mol. Any other form would simply be too long and take up far too much space on a page. Additionally, ignoring the use of scientific notation will make such equations tedious, hard to read, and even harder to remember.

Using Avogadro's number, 1 mol of hydrogen atoms is equal to $6.02214076 \times 10^{23}$. Two mols of hydrogen, then, are equal to $2(6.02214076 \times 10^{23})$.

The number of particles in a mole is the same across elements and substances, but remember not to confuse the *number* with the *mass*. The use of mols also extends into chemical equations, but we will discuss that later on (Encyclopaedia Britannica, 2019b).

Derived Units

Now that we have a grasp on the five primary SI units in chemistry, let us take a brief look at 3 units that have been derived from them: volume, density, and energy.

All substances, apart from those in a gaseous state, have a fixed volume. Volume, as most high school kids will know, is the amount of 3D space a substance or object occupies and varies based on the substance or object's width, length, and height. Volume as a unit is derived from the meter and is calculated by multiplying the object's width by its height by its length (WxHxL).

The methodology for measuring volume will vary between matter states (solid, liquid, or gas) and the 3D space within which they exist. For example, the formula used to find the volume of a square-based pyramid will not be the same as the formula used to find the volume of a sphere.

Square Pyramid Volume Formula

$$\text{Volume} = \frac{b^2 h}{3}$$

Example: Calculate the volume of a pyramid with equal sides = 3cm, height = 4cm

base = 3cm

height = 4cm

volume = ((3²) * 4) / 3

volume = 12cm³

Sphere Volume Formula

$$\text{Volume} = \frac{4}{3} \times \pi r^2$$

Example: Calculate the volume of a sphere with a radius of 3cm

π ≃ 3.14

radius = 3cm

volume ≃ 4/3 * 3.14 * (3²)

volume ≃ 37.70cm³

It is also important to know that when referring to liters, we use cubic centimeters (SI unit symbol: cm^3)

1 cubic centimeter, or 1 cm^3, is equivalent to 1 milliliter.

1,000 cm^3 is equivalent to 1 liter.

1 cubic meter, or 1 m^3, is equivalent to 1,000 liters (Betts, 2016).

Density and volume are closely related. Volume is the 3D capacity that a given object has, whereas the density of an object is the amount of matter taking up the volume. Density is equal to the mass of an object divided by its volume. In mathematical terms, d = M/V.

Energy is an important quantity in many fields of science, including chemistry, physics, biology, etc. Even in the realm of nutrition, energy is an imperative aspect to understand under the concept of calories, or kilocalories in scientific terms.

The term 'calorie' is an older word for the SI unit *joules*. The only difference between the two is that 1 calorie is equal to approximately 4.2 joules (Abozenadah et al., 2017). However, it is best to stick with *joules* as it is more commonly used and recognized.

Energy is most notably defined as the ability or capacity to perform work. In the realm of chemistry, chemical energy can be found in chemical bonds during and after chemical processes and reactions (Abozenadah et al., 2017), in atoms, subatomic particles, and any substance that is used (or has the potential to be used) as fuel. Chemical energy can only be observed and measured *during* a chemical reaction

(Helmenstine, 2020b). There is also ionization energy, but we will be looking into that later on.

For now, let us move on to matter, atoms, and the periodic table. The building blocks of our reality: this time, in a little more detail.

Chapter 2: The Structure of Matter

Atomic Theory

The concept of the atom may seem like a fairly new idea to most. But the first men to conceptualize it lived over 2,000 years ago.

Its origins stem from the pre-Socratic Greeks, where the word atomos (meaning 'indivisible') was first used by the philosopher Leucippus and his pupil, Democritus, in the 5th century BC (Encyclopaedia Britannica & Hosch, 2019). It was later revived by the Roman poet Lucretius, but the Aristotelian view of the elements— that all matter was made up of one or a combination of earth, water, fire, and air— was the most prominent for upwards of 2,000 years (Flowers et al., 2015).

Today, as we have discussed, we understand that an atom consists mostly of empty space, and the majority of its mass is found within the nucleus in the form of protons and neutrons, while a negligible portion of the mass comes from the electrons surrounding the nucleus.

At first, it was believed that atoms were comprised only of electrons existing within a cloud of positive charge that supposedly held them in place. This theory was first proposed by the physicist William Thomson and was strongly supported by Joseph John Thomson, who, just a few years earlier (1897), had discovered the electron. This theory held for a few years before a British physicist named Ernst Rutherford proposed the idea of a nucleus existing at the center of an atom and conducted experiments (which involved the scattering of alpha particles) to disprove the "plum pudding" model of the atom. While ultimately incorrect, Thomson went down in history as the first man to conceptualize the atomic structure (Encyclopedia Britannica & Augustyn, 2017).

At its base, atomic theory is the ancient idea that "all things can be accounted for by innumerable combinations of hard, small, indivisible particles of various sizes but of the same basic material," or, in more definitive and scientifically proven terms, "matter according to which the chemical elements that combine to form the great variety of substances consist themselves of aggregations of similar subunits (atoms) possessing nuclear and electron substructure characteristic of each element" (Encyclopaedia Britannica & Hosch, 2019).

But who was the man who blew Aristotle's theory of air, earth, and fire out of the water? An unlikely genius from an unlikely occupation.

In the year 1807, John Dalton, the man whose name we now use as a unit for measuring atomic mass, was an English schoolteacher when he first proposed his atomic theory and the possibility that all matter could be explained through it.

There are five axiomatic ideas that Dalton's atomic theory brought to the table. Ideas that, while being over 200 years old, are still applicable to modern chemistry.

1. Matter is composed of atoms

This concept seems obvious to us today, but at the time it was a completely new idea. He also realized that all chemical changes take place at the atomic level. What was still to be discovered, however, was that atoms themselves had different parts.

2. A single atom belongs to a single element

One atom cannot have multiple elemental characteristics. Once an atom is identified according to its particular traits, it cannot belong to any other element.

3. One type of element differs from all others at the atomic level

One element cannot share all its properties (proton count in particular) with another.

4. A compound (molecule) consists of two or more elements that are chemically combined

This is, of course, not the same as a mixture. A mixture simply involves combining two or more substances without having them chemically bond.

5. During a reaction no atoms are created nor destroyed (Law of Conservation)

This basic principle reminds us that for a chemical reaction to take place, atoms can neither be created nor destroyed during the process. Instead, they rearrange themselves to change certain properties and produce a new or dissimilar compound or molecule (Flowers et al., 2015).

Lastly, let us take a look at the law of definite proportions and the law of multiple proportions.

The law of definite proportions, coined by the French chemist Joseph Proust, states that "all samples of a pure compound contain the same elements in the same proportion by mass" (Flowers et al., 2015). This concept is fairly straightforward. Imagine you are mixing a bowl of porridge, your favorite kind. You know what *ratio* of milk to porridge you enjoy. Too much more milk, or too little of it, and it's not worth your time. Or, at the very least, it's not as enjoyable. The law of definite proportions is quite similar, if not far more minute in the details.

Your porridge can cope with a few too many drops of water. But a molecule can't. All atoms of a given element must remain within in a precise ratio in a compound for it to remain that molecule.

For example, let's say you have a sample of CO_2 (carbon dioxide, *di-* meaning two, indicating two oxygen atoms for every carbon atom) where a mol of C (carbon) is equal to 12 grams, and a mol of O (oxygen) is equal to 16 grams. (These atomic masses can be easily found using the periodic table— for example, Carbon has an atomic number of 6, but its mass includes protons and neutrons, so the mass = 12)

In your sample of one mol of CO_2, you will have 12 grams of C ($1*12 = 12$) and 32 grams of O ($2*16 = 32$). This ratio can be described as $12/32 = .375$.

Now, if you were to *double* your CO_2 sample, you would simply double your previous numbers according to that same ratio. Double your sample means you will have 24 grams of C ($2*12 = 24$) and 64 grams of O ($2*2*16= 64$). But this ratio of $24/64$ also equals $.375$. Hence, the law of definite proportions.

The law of multiple proportions, on the other hand, is similar while also being distinctly different. The law states that "when two elements combine with each other to form more than one compound, the weights of one element that combine with a fixed weight of the other are in a ratio of small whole numbers" (Encyclopaedia Britannica & Gregersen, 2019).

The first step to observing this law is to have two elements that can combine to form multiple compounds. For this example, we can go back to using carbon and oxygen. C and O can combine to form *multiple compounds*. They are not limited to forming CO_2. Carbon monoxide (*mono-* meaning one oxygen atom), or CO can also be formed.

While CO_2 and CO are made up of the same elements, they are completely different compounds. The atomic mass of carbon, once again, is 12 (remember, atomic number x 2), and the atomic mass of oxygen is 16.

So to find the "ratio of small whole numbers," all we need to do is the following:

Carbon monoxide: $12/16 = .75$

Carbon dioxide: 12/32 = .375

All that's left is to find the ratio between those two numbers: .75/.375 = 2

There will be more examples provided in the 15 end-of-chapter practice questions.

We can now check off the basic ideas that make up the foundation of atomic theory. Next, let us take a proper look at how these different atomic building blocks are arranged in the periodic table. Understanding the principles of how an atom works is one thing, knowing how they are categorized, labeled, organized and named is another equally important aspect of your knowledge to hone.

The Periodic Table

Ah, yes. The dreaded periodic table. The part of chemistry that so many people know of yet so few understand. The numbers, letters, and colors on a periodic table all have a specific meaning that gives us quick insight into the elements within each square. Sort of like the notes in music. A skilled musician needs only a glance at the notes of their song to know what tune comes next. A skilled chemist needs only a glance at their periodic table or a chemical equation to know the names and chemical properties of a given element.

The letters, for example, represent the element's name, either in its abbreviated English name (like helium being shortened to 'He') or its Latin name (like gold being shortened to 'Au' for 'aureus' in Latin).

The first number you will probably notice is that little one in the top left corner of each element square. These indicate the atomic number of each element in the table, also represented by the letter Z. Just remember, Z = Atomic number (Chemistry LibreTexts, 2017); we're going to need that for later. The atomic number is simply the number of protons within an atom's nucleus, which, as has been mentioned, is the manner in which we differentiate elements (Encyclopaedia Britannica, 2019a). For example, hydrogen's atomic number is 1. And as we mentioned earlier, a hydrogen atom has only one proton. The atomic number is also the value we use to organize elements on the periodic table, from the smallest (1, hydrogen) to the largest (118, oganneson).

What about the secondary number on each element in the periodic table? Not all periodic tables display this, but when they do, it refers to the atomic mass of an atom. The atomic mass—often refered to as the atomic weight of an atom—is the combined mass of all the atom's protons, neutrons, and electrons. This would be a whole number if it weren't for the fact that "the value given in the periodic table is an average of the mass of all isotopes of a given element" (Helmenstine, 2020a). We'll discuss isotopes later, but essentially these are exceptions to the atomic rules we've covered so far)

Got the numbers down in your head? Good. Before we get into the details of how the periodic table is organized, let us first look at the differences between gasses, metals, non-metals, and metalloids.

Out of the 3 states of matter—solid, liquid, and gas; we'll ignore plasmagases have the lowest molecular density. They can fit into practically anything, and when placed in a container, will fill it over time through diffusion. Gases behave this way because "their intermolecular forces are relatively weak, so their molecules are constantly moving independently of the other molecules present" (LibreTexts, 2014b). We should remind you at this point that 'molecules' can refer to a combination of elements, like CO_2 or H_2O, but also to pure elements as well, such as Ar, or argon gas.

Metals, on the opposite end of the spectrum, are solids and are not anywhere near as flexible. Metals are classified based on whether or not they have "high electrical and thermal conductivity as well as by malleability, ductility, and high reflectivity of light" (Encyclopedia Britannica et al., 2019). A whopping 75% of the known elements are metals, and of those, aluminum, iron, calcium, sodium, potassium, and magnesium are the most commonly found in the earth's crust.

Non-metals, unlike metals, only display low to moderate conductivity, meaning that some can be used as semiconductors at best and some can be used as insulators. Being non-metals, they differ quite widely from metals, being that they "display a wide range of both mechanical and optical properties, ranging from brittle to plastic and from transparent to opaque" (Encyclopaedia Britannica & Rodriguez, 2019). There are many other qualities that distinguish them from the metals, and non-metals themselves can be distinguished into two groups, covalent materials and ionic materials. We will look at covalent and ionic bonds in the next chapter.

Metalloids essentially fall between metals and nonmetals, playing a sort of hybrid game. When we speak about metalloids, we are typically referring to "simple substance having properties intermediate between those of a typical metal and a typical nonmetal. The term is normally applied to a group of between six and nine elements (boron, silicon, germanium, arsenic, antimony, tellurium, and possibly bismuth, polonium, astatine) found near the center of the P-block or main block of the periodic table" (Encyclopaedia Britannica, 1998). The P-block of the periodic table refers to the right-hand "block" of the table, starting from boron (B, 5) at column 13 ("column" referring to the vertical rows in the periodic table).

Understanding the differences between these groups is not only helpful, it is vital to learn if we are to grasp and utilize the interactions and changes that they go through. Now let us move on to the groups in the periodic table. We will be going from left to right to keep the breakdown coherent.

The first column (or group) in the periodic table consists of what we call alkali metals- lithium (Li, 3), sodium (Na, 11), potassium (K, 19), rubidium (Rb, 37), cesium (Cs, 55), and francium (Fr, 87). While hydrogen is technically in the same column as these metals, it is not part of their group and belongs to the non-metals such as carbon and oxygen. Alkali metals are extremely reactive, meaning that it is nearly impossible for scientists to find pure forms of them in the environment and instead must extract them from compounds found in nature (Averill, 2013). The alkali metals, also known as group IA, are all very reactive, malleable, ductile, and are good conductors of electricity (CrashCourse & Green, 2013).

The second column is made up of alkaline earth metals: beryllium (Be, Atomic no. 4), magnesium (Mg, Atomic no. 12), calcium (Ca, Atomic no. 20), strontium (Sr, Atomic no. 38), barium (Ba, Atomic no. 56), and radium (Ra, Atomic no. 88). These metals are also reactive, though not as aggressively as those in the aforementioned alkali metals. These are known as group IIA.

The middle of the table is made up of transition metals. These have the atomic numbers 21-30; 39-48; 72-80; and 104-108.

The name "transition metals" has no real chemical significance, and aside from them all being metals, most of these elements share the same properties in that they are "hard, strong, and lustrous, have high melting and boiling points, and are good conductors of heat and electricity" (Cotton, 2019). These include iron, nickel, and copper.

Skipping ahead for a moment, the second last column on the table consists of the highly reactive *gases* called halogens. These are fluorine (F, 9), chlorine (Cl, 17), bromine (Br, 35), and iodine (I, 53). These gases react aggressively with alkali metals and alkaline earth metals (CrashCourse & Green, 2013).

Between the halogens and transition metals are a mixed bag of gases, metals, non-metals, and metalloids starting from hydrogen all the way at the beginning of the table, and then going to the following atomic numbers: 5-8; 13-16; 31-34; 49-52; and 81-85.

Then on the far-right column of the table, we have the noble gases: fairly nonreactive elements that are always in a gaseous state at room temperature and, unlike halogens, don't have anger issues.

Lanthanides start on their own island below the main table. Their group includes elements with the atomic numbers 58-71, as well as the namesake element 57, lanthanum. These are all highly reactive metals with high melting and boiling points that have a silver coloration and tarnish when oxidized.

And lastly, we have the row below the lanthanides, the radioactive actinides like uranium and plutonium, made up of the elements with the atomic numbers 90-103.

Periodic Trends

Next, let us take a look at periodic trends. The main periodic trends are "electronegativity, ionization energy, electron affinity, atomic radius, melting point, and metallic character" (Chemistry LibreTexts, 2013). Periodic trends are a convenient way for chemists to quickly find and predict the properties of an element.

To start, let's define Coulomb's law. The law is named after French physicist Charles-Augustin de Coulomb and states that:

- The force between two electrical charges is proportional to the product of the charges and inversely proportional to the square of the distance between them. (Encyclopaedia Britannica, 2021).

- Like charges repel each other; unlike charges attract. Thus, two negative charges repel one another, while a positive charge attracts a negative charge.
- The attraction or repulsion acts along the line between the two charges.
-

Understanding this law will help you easily grasp all of the periodic trends. Remember this unit of measurement as well: the coulomb. Coulomb's law also incorporates the concept of electrostatic forces, also known as Coulomb forces or Coulomb interactions. These refer to the "attraction or repulsion of particles or objects because of their electric charge" (The Editors of Encyclopaedia Britannica & Gregersen, 2016).

The magnitude of this force can be calculated using the equation $F = \frac{kq_1q_2}{r^2}$.

F here stands for the electric force, 'k' is the proportionality constant, 'q_1' is the electric charge of the first particle, and 'q_2' is the electric charge of the second particle). When using SI units of force (F, in Newtons), charge (C, in Coulombs), and distance (m, in meters), the proportionality constant k = 8.99 x 10^9.

Electronegativity Trends

Electronegativity is the "ability of an atom to attract to itself an electron pair shared with another atom in a chemical bond" (The Editors of Encyclopedia Britannica, 2011).

Basically, it's the measure used to determine how badly an atom wants to gain an electron. The primary scale used for quantifying electronegativity is the Pauling scale, named after a chemist named Linus Pauling. Electronegative values are displayed below each element in some periodic tables (Chemistry LibreTexts, 2013).

Most atoms follow the octet law (the outer valence shell or orbital containing 8 electrons). However, because "elements on the left side of the periodic table have less than a half-full valence shell, the energy required to gain electrons is significantly higher compared with the energy required to lose electrons. As a result, the elements on the left side of the periodic table generally lose electrons when forming bonds" (Chemistry LibreTexts, 2013). We will be discussing the different types of bonds in the next chapter.

Electronegativity tends to increase when moving from left to right and decreases moving from top to bottom for *most* columns—emphasis on the word 'most'. Noble gases, lanthanides, and actinides are all groups whose rows/columns don't follow the aforementioned trends. Transition metals do have electronegativity values, but due to their metallic properties, the values themselves do not have much variability across the group.

Ionization Energy Trends

Ionization energy (measured using electron volts, eV, or kilojoules per mole, kJ/M) is "the energy required to remove an electron from a neutral atom in order to turn it into an ion" (Helmenstine, 2020d). An ion is therefore an atom that has a missing or an extra electron.

Similar to electronegativity, ionization energy tends to increase when moving from left to right across the periods (or rows) of the periodic table, and decreases moving top-down the element groups (or columns). The higher an element's ionization energy, the easier it is for it to become an anion, meaning it has a greater negative charge than positive charge having gained an electron. Logically enough, the inverse is also true; the lower an element's ionization energy, the more likely it is to become a cation. We will take a closer look at cations (+) and anions (-) in the next chapter.

Some elements have multiple ionization energies, and these are most commonly called "the first ionization energy, the second ionization energy, third ionization energy, etc" (Chemistry LibreTexts, 2013). "To remove the outermost valence electron from a neutral atom is the first ionization energy. The second ionization energy is that required to remove the next electron, and so on" (Helmenstine, 2020d).

The first electrons to be removed will be those in the outermost shell, called valence electrons. In general, the closer the electron is to the nucleus, the higher its ionization energy will be.

When looking at the ionization energy of any element, simply remember that it is referring to the amount of energy (normally measured in kj/mol) required to remove the *first electron* (one of the valence electrons) from the given atom to turn said atom into a positively charged ion, or cation.

Electron Affinity Trends

"Electron affinity is defined as the change in energy (in kJ/mole) of a neutral atom (in the gaseous phase) when an electron is added to the atom to form a negative ion. In other words, the neutral atom's likelihood of gaining an electron" (Chemistry LibreTexts, 2019b). While we use ionization energies for the formation of positive ions, electron affinities are used for negative ions (anions). In addition to that, it is useful to know that "their use is almost always confined to elements in groups 16 and 17 of the Periodic Table" (Chemistry LibreTexts, 2019b). The trend of electron affinity is the same as ionization energy and electronegativity. It increases moving upward and from left to right across periods. This is because "the electrons added to energy levels become closer to the nucleus, thus a stronger attraction between the nucleus and its electrons" is formed (Chemistry LibreTexts, 2019b). This is Coulomb's inverse square law in action; much like magnets (or gravity) the greater the distance, the weaker the attraction.

As we are seeing so far, many of these trends correlate. In the next section, atomic radius trends, notice how it has a direct correlation and *causation* with electron affinity. We just mentioned how the electron affinity of an atom is affected by the distance between the nucleus and electrons. Now let us take a look at the trends of atomic radius, and see if you can spot the correlation yourself.

Atomic Radius Trends

Atomic radius (or radii, plural), is "one-half the distance between the nuclei of two atoms, just like a radius is half the diameter of a circle" (Chemistry LibreTexts, 2013).

Due to the fact that not all atoms are typically found bound together, this idea is somewhat complicated. Some molecules are affiliated by covalent bonds, while others are attracted to each other in ionic crystals, while others are held together in metallic crystals. Despite this, a large majority of elements can form covalent molecules, in which two atoms of like type are joined together by a single covalent bond. These molecules' covalent radii are often called their atomic radii. This distance is measured in picometers, and the patterns of the atomic radius can be observed in all parts of the periodic table (Chemistry LibreTexts, 2013).

Unlike the other trends in the periodic table, atomic size decreases from left to right across a period of elements as opposed to increasing in electron affinity, for example. The reason for this decreasing trend is that all electrons in a family or period of elements are assigned to the same shell.

The nucleus, however, is becoming more positively charged at the same time as protons are added. Remember that protons are around 1800x the size of electrons, and so an increase in the number of protons has a greater effect than increasing the number of electrons, meaning there is a stronger nuclear attraction. Consequently, a higher proton count means the electrons are more attracted to the nucleus, pulling the shell closer to it (Chemistry LibreTexts, 2013). Think of the valence electrons as a python, and the nucleus of the atom as a rat being crushed inside its coils. The tighter the valence electrons (the python), the closer it is to the nucleus of the atom (the rat). Essentially, the atomic radius decreases as the valence electrons draw closer.

Melting Point Trends

An element's melting point is the energy required to change its state from its solid to its liquid state. This conversion between states requires the breaking of bonds. Thus, the stronger the bond between the atoms of a given element, the higher that element's melting point will be.

A high bond dissociation energy correlates with a high temperature, as temperature is directly proportional to energy. Melting points, however, vary and do not generally follow a traceable trend.

There are some rules to bear in mind when considering or predicting an element's melting point. "Metals generally possess a high melting point. Most non-metals possess low melting points. The non-metal carbon possesses the highest melting point of all the elements. The semi-metal boron also possesses a high melting point" (Chemistry LibreTexts, 2013).

Metallic character refers to the set of chemical properties found in elements that are metals. These chemical properties can be attributed to the ease with which metals lose electrons to form cations (positively charged ions).

As you move across and down the periodic table, there is a trend in metallic character. Metallic character increases as you move to the center and lower rows of the periodic table. The reason for this is that they have valence electrons in the two outer shells. This means that electrons in both shells are involved in chemical bonding, making bonds much stronger.

The number of metallic characteristics in elements increases as you move down an element group (vertical). This is in correlation with atomic radius trends—as the atomic radius increases, electrons become easier to lose since there is less attraction between the nucleus and the valence electrons.

That brings us to the end of the periodic trends. It is important to understand how the periodic table is organized and labeled so that you can become more proficient at using it. The deeper you decide to go into the world of chemistry, the more useful these trends will become, particularly when you move off the theoretical and into the practical.

It is also interesting to see how many of these trends are correlated. One trend is often indicative of the other, meaning if you have the value(s) of one or more, you can calculate the value of another trend depending on how well you understand those trends.

The following questions are designed to recap the information we have learned as well as test your understanding of them.

Practice Questions

You will need a periodic table at hand to answer these questions. Remember the periodic trends and the various laws we have discussed in this chapter.

1. **Question: Define the law of definite proportions.**

   Answer: All samples of a pure compound contain the same elements in the same proportion by mass.

2. **Question: Use the law of definite proportions to calculate the following: If 3.5 g of element X reacts with 10.5g of element Y to form the compound XY, how many grams of Y would be needed to form XY2?**

   Answer: Y2 = 2 (Mass of Y)

   $$= 2 \,(10.5g)$$

   $$= 21.0 \text{ g}$$

3. **Question: Define the law of multiple proportions.**

   Answer: All samples of a pure compound contain the same elements in the same proportion by mass.

4. **Question: Nitrogen forms multiple compounds with oxygen. Measurements of the masses of nitrogen and oxygen that form upon decomposing these compounds indicate**

that one compound contains 2.28g of oxygen per 1g of nitrogen, while the other compounds contain 0.057g oxygen for every 1g of nitrogen. Calculate the simplest whole-number ratio using the law of multiple proportions.

Answer: Compound 1: 2.28/1 = 2.28

Compound 2: 0.057/1 = 0.057

Ratio: 2.28/0.057 = 40/1

5. **Question: Two compounds containing phosphorus and chlorine are shown below as mass composition values. Show why these values follow the law of multiple proportions.**

Substance	Mass of Phosphorus (g)	Mass of Chlorine (g)
Compound 1	3.097	10.63
Compound 2	1.548	8.863

Answer: (10.63/3.097) ÷ (8.863/1.548) = 0.6 = 3/5

B. Question- If Compound 1 above was found to have a formula of PCl3, propose a reasonable formula for Compound 2.

Answer: Assuming a fixed amount of phosphorus, the ratio of Compound 1 to Compound 2 is 3:5. So if Compound 1 is PCl3, then Compound 2 could be PCl5.

6. Question: Which atom is the most electronegative?

Silicon

Aluminium

Chlorine

Sulfur

Phosphorus

Answer: Chlorine.

Explanation:

Electronegativity refers to the ability of an atom to attract and bind electrons. Since Cl has the highest number of protons in this group of atoms, the electromagnetic pull on its electrons is the highest, reducing the atomic radius and increasing the attraction for an extra electron to be pulled in. Once that happens, the outer shell of the Cl atom becomes full (8 electrons), and the atom becomes the ion Cl$^-$. This explains why Cl$^-$ anions are so attracted to sodium ions (Na$^+$), resulting in the molecular compound NaCl (table salt).

7. **Question: Which of the following elements has the most metallic character:**

Bi

P

Sb

As

Answer: Bi

Explanation:

As you move down in the periodic table, metallic characteristics become more prevalent. Since Bi is the lowest in this Group, it must have the most metallic character according to the metallic character periodic trend.

8. **Question: Which of these atoms is the smallest?**

Calcium

Strontium

Barium

Beryllium

Magnesium

Answer: Beryllium

Explanation:

Each row in the periodic table represents a new layer of orbitals, or shells. Since beryllium is on the second row, it has only two shells, its outer shell containing two electrons. All the other atoms in this group have higher numbers of shells, increasing the size of the atoms as you go down the periodic table.

9. **Question: Which group of elements has the highest first ionization energies?**

   Alkaline earth metals

   Noble gases

   Alkali metals

   Halogens

   Metalloids

   Answer: Noble Gases

Explanation:

Ionization energy refers to the difficulty or energy required to remove an electron from the outer shell of an atom. The noble gases are generally inert (non-reactive and very resistant to losing or gaining an electron) because their outer shells are full. There are eight electrons (octet rule) in the outer shell of each of the noble gases (aside from helium, which has two). The highest electron affinities are found in the halogens, which is the reason why they have high—but not the highest—ionization energies. Metalloids have ionization energies that lie more in the middle of the two extremities. Alkali and alkaline earth metals have low ionization energies (indeed, these atoms are happy to donate an extra electron or two) and thus have low electron affinities (Varsity Tutors, n.d. -c).

10. **Question: Which of the given atoms has the lowest electron affinity?**

Sr

Ca

Ra

Be

Answer: Ra

Explanation:

The group of alkaline earth metals includes beryllium, calcium, strontium, and radium. The electron affinity, or the amount of energy released when an atom gains an electron (an exothermic reaction), decreases from the top of a column to the bottom.

"Correlations can be found between electron affinity and ionization energy. When a smaller atom gains an electron, the force between the electron and nucleus is greater than in a larger atom; thus, more energy is released when this 'bond' between the nucleus and electron is formed in a smaller atom than in a larger atom, meaning that smaller atoms will have greater electron affinity. Radium is the farthest down the group of alkaline earth metals and will have the largest atomic radius of the answer choices, giving it the lowest electron affinity" (Varsity Tutors, n.d.-a).

11. **Question: Which of the following atoms is the largest?**

Phosphorous

Sulfur

Sodium

Chlorine

Aluminum

Answer: Sodium

Explanation:

"Sodium, aluminum, phosphorus, sulfur, and chlorine are all in the same row of the periodic table. Since the size of the atomic radii decreases as you move from left to right on the periodic table, the element furthest to the left will be the largest. Therefore, sodium is the largest atom in the group" (Varsity Tutors, n.d.-c).

12. **Question: What is Coulomb's law?**

Answer: Charges of the same polarity repel each other, while charges of opposite polarity attract each other, with a force proportional to the product of the charges and inversely proportional to the square of the distance between them (OWTTE).

13. **Question: What are the 5 main ideas stated by John Dalton's atomic theory?**

Answer:

1. Matter is composed of atoms. All chemical changes take place at the atomic level.

2. A single atom can only contain the chemical characteristics of a single element. One atom cannot have multiple elemental characteristics. Once an atom is identified according to its particular traits, it cannot belong to any other element.

3. One type of element differs from all others at the atomic level. An element cannot share all properties (e.g, proton count) with another.

4. A compound consists of two or more elements that are chemically combined.

5. During a reaction, no atoms are created nor destroyed (Law of Conservation of Mass)

14. **Question: How are electrons and protons similar, and how do they differ?**

Answer: "Electrons and protons are both charged subatomic particles. Protons are much larger than electrons (contributing more mass to the overall atom). Changing the number of protons changes the identity of the atom while changing the number of electrons changes the charge" (LibreTexts, 2020).

15. **Question: "Which postulate of Dalton's theory is consistent with the following observation concerning the weights of reactants and products? When 100 grams of solid calcium carbonate is heated, 44 g of CO_2**

and 56 g of CaO are produced" (LibreTexts, 2020).

Answer: Chemical changes do not create or destroy atoms, but instead reorganize them to produce substances that differ from the ones already present. This can be inferred from the law of conservation of mass.

Chapter 3: Chemical Bonds

Have you ever heard the saying, opposites attract? Well, that sentence is questionable with regard to human relationships, and in many ways this saying is not always true with regard to chemical bonds, either.

But first, before we can delve into chemical bonds, it's important to understand the organization of electrons within an atom. Electron configuration will help you distinguish the different bonding types with a little more ease.

Electron Configuration

Electron configuration, or the way in which we represent how electrons are arranged within orbitals and subshells, is most commonly used to describe the orbitals of an atom in its ground state (Faizi & Chemistry LibreTexts, 2016).

However, it can also be used to represent an atom that has been ionized—added or subtracted electrons to create an anion or cation— by taking into account electrons lost or gained in subsequent orbitals. There are many elemental properties, both physical and chemical, that can be linked to the said element's unique electron configurations (Faizi & Chemistry LibreTexts, 2016).

The chemistry for each element in the periodic table is unique to that element and is determined by the valence electrons (the electrons in the outermost shell).

All electrons are located around the nucleus of an atom within the atom's electron orbitals, which can be defined as "the volume of space in which the electron can be found within 95% probability" (Faizi & Chemistry LibreTexts, 2016).

The four main orbitals are s, p, d, and f. An easy way to remember them is to link a series of strangely connected words to the letters. SPDF— Space Patrol Dolphins Flying. Or something like that. Make sure it's an absurd abbreviation so that you remember it better. SPDF actually stands for sharp, principal, diffuse, and fundamental, but remembering the letters is what's important for now at least.

The number of electrons held in each shell (and/or subshell) is what we're going to take a look at first. We know from our section on the periodic table that the atomic number of an element (the number in the top left corner on each element in the periodic table) is equivalent to the number of protons in an *atom*. At its neutral state, this is also the number of electrons, but this number can change when the atom loses or gains an electron to become an ion.

When dealing with the electron configuration of atoms, it is quite simple to understand and use the SPDF shells.

When an atom gains an electron, it occupies an orbital in order to minimize the energy of the atom. With there being multiple orbitals, how do we know which orbital an electron goes to? Simply put, electrons in an atom fill the orbitals, or energy levels, in order of increasing energy as the electrons get farther away from the nucleus (Faizi & Chemistry LibreTexts, 2016). If you add an electron to an atom, it will always be added to its valence shell.

The order in which these energy levels are filled looks like this:

1s, 2s, 2p, 3s, 3p, 4s, 3d, 4p, 5s, 4d, 5p, 6s, 4f, 5d, 6p, 7s, 5f, 6d, and 7p (Faizi & Chemistry LibreTexts, 2016). 1s has the lowest energy level, and 7p has the highest.

This may seem daunting to remember, but again, it is quite simple once you know the pattern. Don't think about these orders linearly, in the way they are written above. Think about them like so:

1s

2s 2p

3s 3p 3d

4s 4p 4d 4f

5s 5p 5d 5f

6s 6p 6d

7s 7p

We will call this the Diagram of Orbital Order (DOO), though you will likely come across other names for it in the future.

The tricky part here is training yourself to think of these numbers *diagonally*. Perhaps remembering them linearly sounds easier, but so long as you have a pen and paper (or photographic memory), you can easily identify the order.

First, remember SPDF. Then remember the numbers 1-7.

The first row only has 1 orbital—1s (the *first letter* of SPDF).

The second row has 2—2s and 2p (the first *2 letters* of SPDF).

The third row has 3—3s, 3p, and 3d (the first *3 letters* of SPDF).

The fourth row has 4—4s, 4p, 4d, and 4f (all *4 letters* of SPDF).

And then we start counting backward. The fifth row has the same layout as row 4.

The sixth row has one less—6s, 6p and 6d (what helps me remember this is that 6 divided by 2 is 3, the amount of orbitals in this row, s, p and d).

And the seventh row has two once again—7s and 7p.

Let's put this in a table:

$1s^2$			
$2s^2$	$2p^6$		

$3s^2$	$3p^6$	$3d^{10}$	
$4s^2$	$4p^6$	$4d^{10}$	$4f^{14}$
$5s^2$	$5p^6$	$5d^{10}$	$5f^{14}$
$6s^2$	$6p^6$	$6d^{10}$	
$7s^2$	$7p^6$		

The numbers in superscript are the maximum number of electrons that can fit in each orbital.

Once you write out the progression of orbitals this way, you can move on to the next step. Simply draw a diagonal line through each number, crossing the squares downward and to the left, and you have the order of energy levels.

Orbital s can hold 2 electrons, but p can hold 6 electrons, d can hold 10 electrons, and f can hold 14 electrons (Helmenstine, 2020c).

Let's take a look at an example of how you would write the electron configuration of an atom. Nitrogen (N) has an atomic number of 7. Thus, it has both 7 protons and 7 electrons. To write nitrogen's electron configuration, we have to write out the energy levels in the order in which they are filled (1s, 2s, 2p, 3s, 3p, 4s, 3d, etc.) until the number of electrons reaches 7.

Remember that s orbitals can only hold 2 electrons, and p orbitals can hold up to 6, including their subshells.

Thus, we write nitrogen's electron configuration as $1s^2$ $2s^2$ $2p^3$. The superscripted numbers all refer to the number of electrons in that specific orbital. As you can see, the numbers add up to 7 electrons. Not every orbital has to be filled to its max, which is why 2p, despite being able to hold up to 6 electrons, only holds 3 ($2p^3$) for this particular element.

As long as you get the order of the orbitals, the maximum number of electrons that each orbital can hold, and the sum of the total values to match the element's atomic number correctly, you will do just fine.

Shorthand Electron Configuration

Also called core-abbreviated electron configurations, shorthand electron configurations "replace the core electrons with the noble gas symbol whose configuration matches the core electron configuration of the other element" (Flowers et al., 2018).

This, once again, is very simple. Every element exists within a period (horizontal row) on the periodic table. Every period ends with one of the 7 noble gases (helium He, neon Ne, argon Ar, krypton Kr, xenon Xe, radon Rn, and oganesson Og). Every noble gas has a full electron configuration.

This will sound very much like a periodic trend, but every element has "core electrons" that are equivalent to the electron configuration of the noble gas in the *previous* period. A shorthand configuration, therefore, is the symbol of that noble gas *plus* the element's additional electrons in the form of an electron configuration.

For example, the shorthand configuration of calcium is [Ar]4s^2. Why?

Calcium exists within the fourth period of the periodic table, and so argon (Ar) is the noble gas at the end of the *previous period* (the third period).

Calcium's full electron configuration is 1s^2 2s^2 2p^6 3s^2 3p^6 4s^2.

Argon's is 1s^2 2s^2 2p^6 3s^2 3p^6.

Notice that calcium's first 5 orbitals are the same as argon's. So, instead of writing calcium's entire electron configuration, we can write the shorter version of [Ar]4s^2. This can also help you find the atom's valence electrons, as the electrons 4s^2, for example, are calcium's valence electrons. The rest of them (or those that are the same as argon's) are its core electrons.

The Aufbau Principle

Aufbau comes from the German word "Aufbauen" which means "to build". As we progress from atom to atom, we build up electron orbitals when writing electron configurations. We will fill the orbitals of an atom in increasing order of atomic number when we write down the electron configuration for an atom (Chemistry LibreTexts, 2019a).

"The Aufbau principle originates from Pauli's exclusion principle, which says that no two fermions (electrons) in an atom can have the same set of quantum numbers, hence they have to "pile up" or "build up" into higher energy levels. How the electrons build up is a topic of electron configurations" (Chemistry LibreTexts, 2019a).

Aufbaus' principle acts more like a rule in the sense that it has its exceptions. The d and f block elements do not always follow this principle because half-filled or completely filled subshells add stability to atoms. For example, the predicted Aufbau configuration for chromium (Cr, Atomic No. 24) is $4s^2$ $3d^4$, but the actual configuration is $4s^1$ $3d^5$ (Helmenstine, 2020e).

The Two Main Bonds

Opposites attract in chemistry and create strong bonds, which is rarely the case with human beings. Humans form bonds with people who have similar tastes, values, morals, and world views. Atoms, on the other hand, often form bonds with other atoms that are completely opposite to themselves in terms of electrical charge.

Two atoms or ions bond in order to lower their energy. But how do these bonds occur? Let's say, for example, that you want to find out how a sodium (Na, Atomic No. 11) atom combines with a chlorine (Cl, Atomic No. 17) atom to create sodium chloride (NaCl, also known as table salt). Remember that neutrons have no charge, while electrons are negatively charged and protons are positively charged.

All atoms in a neutral state have the same number of protons as they do electrons, meaning that they are uncharged due to the fact that this balance of charged particles cancels the charges out. In our example, sodium has an atomic number of 11 and thus has 11 protons and 11 electrons in a neutral state. Chlorine, on the other hand, has an atomic number of 17 and thus has 17 protons and 17 electrons. However, when sodium and chlorine bond, the sodium atom (a metal) gives one of its electrons to the chlorine atom (a non-metal) to bond the two together.

When this happens, sodium loses an electron and becomes positively charged (as the atom now has one more proton than it has electrons), and the chlorine atom becomes negatively charged due to gaining an electron (because the electron is a negatively charged particle, and the chlorine atom that receives it now has one more electron than it has protons). These charged atoms are known as ions. A positively charged ion (in this case, sodium) is called a cation, and a negatively charged ion (in this case, chlorine) is called an anion.

These bonds have names based upon property-driven nomenclature that we will be discussing throughout this chapter.

Ionic bonds and covalent bonds are the two types of chemical bonds, and we can also separate covalent bonds into nonpolar covalent bonds and polar covalent bonds. We will also be looking at the practical use of Coulomb's law in measuring the interactions between ions. All these topics and more are what we will be looking at for the rest of this chapter.

Ionic Bonds

Ionic bonds are a "type of linkage formed from the electrostatic attraction between oppositely charged ions in a chemical compound. Such a bond forms when the valence (outermost) electrons of one atom are transferred permanently to another atom" (The Editors of Encyclopedia Britannica, 2014).

In ionic bonds, one atom gives an electron to stabilize another atom and form a bond with it. Another way of looking at it is that the donated electron spends most of its time close to the bonded atom (Helmenstine, 2021a).

Ionic bonds can only occur between atoms that have different electronegativity values from each other.

In polar bonds, oppositely charged ions are attracted to each other. An example of this is that of the NaCl bond we mentioned earlier.

"...sodium and chloride form an ionic bond, to make NaCl, or table salt. You can predict an ionic bond will form when two atoms have different electronegativity values and detect an ionic compound by its properties, including a tendency to dissociate into ions in water" (Helmenstine, 2021a).

When the valence shell of an atom isn't full, it becomes, in a sense, angry. All atoms want to have a full valence shell, so when an atom has a valence shell that is only partially full, it becomes unstable.

This is where the Octet rule comes into play once again. Though this rule does not apply to all atoms, many atoms that have 8 electrons in their valence shell are the most stable. Atoms that meet the Octet rule have an electronic configuration that is similar to those of noble gases (ScienceABC, 2020).

Now you may be wondering, why does the electron move from the sodium atom to the chlorine atom and not the other way around?

To answer that question, we will have to go back to electronegativity trends for a moment. Because sodium is a metal, it has a lower electronegativity (meaning it has less interest in gaining an electron) than chlorine, which is a non-metal and has a relatively high electronegativity (meaning it is very interested in gaining an electron). Metals tend to have lower electronegativity than nonmetals, so when dealing with metal to nonmetal ionic bonds, the metal will always lose an electron and the non-metal will always gain it (LibreTexts, 2019). This also means that the metal atom will always become positively charged (net charge of +1 if it loses 1 electron, meaning it becomes a cation) and the non-metal will always become negatively charged (net charge of -1 if it gains 1 electron, meaning it becomes an anion).

Molecules that have ionic bonds clump together to form what are known as lattice structures. There are many kinds of lattice structures, but for now, we will only be going over the simple cubic structure.

This structure can be conceptualized as a group of ping-pong balls (or some other spherical object) placed within a cube container in layers that allow for the balls to be connected. The second simplest way to understand it is with a series of ionically bonded atoms (for example, NaCl) stacked across each other and interconnected (each Na atom connected to four Cl atoms in a 2D depiction) to form a cube made up of spheres (atoms).

Lattice structures in the form of crystalline solids account for around "90% of naturally occurring and man-made solids... Most solids form with a regular arrangement of their particles because the overall attractive interactions between particles are maximized, and the total intermolecular energy is minimized, when the particles pack in the most efficient manner. The regular arrangement at an atomic level is often reflected at a macroscopic level" (OpenTextBC, 2019).

Covalent Bonding

Covalent (or molecular) bonding is the more widely used bonding type. It is the bond formed between atoms of the same element or of those close to each other on the periodic table.

This bond, like ionic bonds, involves the sharing of electrons between atoms. Unlike ionic bonds, however, it is not limited to occurring between metals and nonmetals. While this bond type primarily occurs between nonmetals, it can also be observed between metals and non-metals (LibreTexts, 2019).

Atoms with similar electronegativities (the same hunger for electrons) are most likely to form covalent bonds. Due to their similar affinity for electrons, and because neither atom tends to donate them, the two atoms share electrons to achieve an octet configuration similar to that of noble gases, and become more stable. Furthermore, the ionization energy of the atom is too high and the electron affinity of the atom is too low for it to form an ionic bond.

Carbon, for instance, does not form ionic bonds since it has four electrons, or half an octet. Thus, the carbon molecule must gain four electrons (to fill its valence shell by following the octet rule) or lose four electrons (empty its valence shell and, in essence, shed it so that its next shell down becomes its *full* valence shell) in order to form an ionic bond. That is not the optimal route, and as such, carbon atoms will share their 4 valence electrons by way of single, double, and triple bonds, resulting in noble gas configurations for every atom they bond with (LibreTexts, 2019).

In other words, in covalent bonds, electrons are not donated or given away as they are in ionic bonds, but are rather shared. They are kept by their original atom and bond to another atom as well. Take fluorine, for example. Fluorine only has 7 electrons in its valence shell. Due to it being a halogen, it is highly reactive and so fluorine atoms often travel in covalently bonded pairs. When two of these atoms bond, as always, they are trying to fill their valence shell and have 8 electrons.

To do this, one cannot donate an electron to the other, as this will only leave one of the fluorine atoms with an octet of electrons in its valence shell. Instead, the two will bond and share an atom each, meaning both will, in a sense, gain an electron and fill their valence shell. Sharing electrons without losing them can also be conceptualized as the orbitals (and thus the electrons) overlapping without having an electron actually leave either of the involved orbitals.

Here's a more detailed way of looking at it using electron dot structures, also known as the Lewis electron dot structure.

"The diatomic fluorine molecule (F2) contains a single shared pair of electrons. Each F atom also has 3 pairs of electrons that are not shared with the other atom. A lone pair is a pair of electrons in a Lewis electron dot structure that is not shared between atoms. Each F atom has 3 lone pairs. Combined with the two electrons in the covalent bond, each F atom follows the octet rule" (LibreText, 2016).

What are single, double, and triple bonds?

Single bonds are nothing more than a single covalent bond between the atoms in a molecule. As quoted earlier, a fluorine atom with 7 electrons will only share 1 pair of electrons with its molecular partner and leave the other 3 pairs alone or unbonded. This is, of course, a simplified way of looking at single bonds using the Lewis electron dot structure of covalent bonds, but it is an adequate model that somewhat accurately depicts how these bonds work.

To meet the octet rule, however, some molecules cannot have only one covalent bond between their atoms. To rectify this issue, double and triple covalent bonds are formed. A double covalent bond is formed by atoms that share two pairs of electrons.

A triple covalent bond is a covalent bond formed by atoms that share 3 pairs of electrons. Take the element nitrogen for example. Nitrogen is a gas that composes most of the Earth's atmosphere, making up around 78% of it (NASA Global Climate Change, 2016). A nitrogen atom has five valence electrons, which, in a Lewis electron dot structure, can be depicted as one pair of electrons and 3 single electrons in the valence (outer) shell. The 3 single electrons of a nitrogen atom combine with those of another nitrogen atom to form 3 shared pairs of electrons, creating what we call a diatomic molecule. Diatomic elements are never found alone, and are instead found in pairs of two as molecules. The diatomic elements are Br2, I2, N2, Cl2, H2, O2 and F2.

Additional Information & Other Bonds

The Difference Between Ionic And Covalent Bonds

Ionic bonds (or ionic compounds) occur between metals and nonmetals and the bond is formed when one atom, in a sense, steals an electron from the other. The change in charge between the atoms as a consequence of this is what bonds the two together. This connection is similar to that which we observe between magnets.

Covalent bonds (or molecular compounds) occur between two nonmetals and are held together by the atoms *sharing* electrons. Neither atom loses or gains an electron in the same way that the atoms in an ionic bond do. Instead, their valence shells overlap to share their electrons, sometimes by means of multiple covalent bonds, in order to fill their valence shells.

Here's another way in which the two differ. When dissolved in water (H_2O), ionic bonds dissociate from a lattice structure into individual atoms. Covalent bonds, on the other hand, break apart at the molecular level. They do *not* break down into individual atoms (Helmenstine, 2021a).

Polar and Nonpolar Covalent Bonds

Polar bonds form when the electronegativity difference between the anion and cation is between 0.4 and 1.7 (Helmenstine, 2021b). The electrons forming the bond between two atoms must also be distributed unequally for them to be classed as polar. Remember; polar bonds have an unequal electron distribution, and non-polar bonds have an *equal* electron distribution.

To find out what the relative polarity of a covalent bond is, chemists use electronegativity values, which, as we have discussed, "is a relative measure of how strongly an atom attracts electrons when it forms a covalent bond" (Chemistry LibreTexts, 2016b).

A polar covalent bond is formed between nonmetal atoms when their electronegativities are sufficiently different (0.4-1.7, though these values may differ according to different sources). The inequality in electronegativity values is what causes the bonding electron pair to be unequally distributed between atoms. Hydrogen, for example, forms polar covalent bonds with any nonmetal. The electron distribution of polar bonds will cause one end of the molecule to have what is called an electrical dipole moment, whereas the other end has a slight negative dipole moment. Molecules with polar covalent bonds interact with other molecules' dipoles because this type of bond separates positive and negative charges. This produces dipole-dipole intermolecular forces between the molecules (Helmenstine, 2021b).

In molecules, dipoles are caused by unequal electron sharing between atoms. The electrons bonded to more electronegative atoms are pulled closer to the atom itself. In molecular dipoles, one side of a molecule has a partially negative charge and the other side has a partially positive charge as a result of electron density building up around an atom or discrete region of a molecule (Bertrand et al., 2013). A dipole-dipole force is an attraction between the positive end of one polar molecule and the negative end of another polar molecule. A mole's dipole-dipole force is between 5 kJ and 20 kJ. Unlike ionic or covalent bonds, they are much weaker and do not form bonds between atoms. Like magnets, they only have a significant effect when the molecules involved are almost close enough to one another to be touching (Purdue University Department of Chemistry, n.d.).

Polar bonds divide pure covalent bonds from pure ionic bonds (Helmenstine, 2021b).

All covalent bonds between atoms belonging to different elements are classified as being polar, though the degree of polarity varies from bond to bond. Some bonds only have minimal polarity while others are strongly polarized. Ionic bonds have what can be called the "ultimate in polarity, with electrons being transferred rather than shared" (Chemistry LibreTexts, 2016b).

A nonpolar bond occurs when the electrons are *equally* distributed between the atoms making up a molecule.

Non-polar molecules can easily be identified when a covalent bond forms between atoms that have the same or similar electronegativity values. When molecules share electrons *equally* in a covalent bond, no net electrical charge is generated within the molecule. Broadly speaking, when the electronegativity difference between two atoms is *less* than 0.5, the bond is nonpolar. This is why knowing electronegativity trends in the periodic table is so useful. However, it is important to note that only molecules formed by two identical atoms are truly non-polar (Helmenstine, 2020f).

This is because non-polarity requires the closest similarity in electronegativity values, and, of course, the two atoms with the closest electronegativity values will always be those of the same element.

For example, in the molecule H—H (H_2), both hydrogen atoms have electronegativity values of 2.2, and thus the bond is considered non-polar.

Another way of looking at this is that non-polar molecules form when atoms sharing a polar bond are arranged such that the electric charges cancel each other out.

Knowing the polarity of different molecules enables you to predict whether or not they will form chemical solutions when mixed together.

Remember that homogeneously polar/non-polar molecules dissolve. In other words, "polar molecules will dissolve into other polar liquids and nonpolar molecules will dissolve into nonpolar liquids" (Helmenstine, 2020f).

This is why trying to wash grease or oil or butter off of your hands with water doesn't work. The molecules making up those mixtures are non-polar, whereas water molecules are polar, which means that the two repel each other, so water is practically useless at dissolving and/or rinsing off those substances.

The idea of knowing which chemicals are intermediate between polar and nonpolar is extremely useful because you can use them as an intermediary to dissolve a chemical into one it wouldn't mix with otherwise. It may be possible to dissolve an ionic or polar compound in ethanol (mildly polar) if you are mixing it with an organic solvent. An organic solvent like xylene can then be used to dissolve the ethanol solution (Helmenstine, 2020f).

Sigma and Pi Bonds

When we talk about sigma and pi bonds, we are simply referring to the varieties of covalent bonds that occur between atoms. The way we look at electrons interacting can be depicted as nothing more than overlapping circles, but when we start getting into double and triple covalent bonds, that simple diagram no longer works. We need a more complex view of how these electrons interact to get a more accurate understanding of them (Chemistry LibreTexts & Editors, 2016). These more complex structures are known as hybridization models.

Before we go any further, let us take a brief look at sigma and pi bonds.

"A pi bond (π bond) is a covalent bond formed between two neighboring atoms' unbonded p-orbitals. An unbound p-orbital electron in one atom forms an electron pair with a neighboring atom's unbound, parallel p-orbital electron. This electron pair forms the pi bond" (Helmenstine, 2019c). Pi bonds cannot exist independently of sigma bonds.

Single bonds between atoms are made up of a single sigma bond.

Double bonds between atoms are made up of 1 sigma bond and 1 pi bond.

Triple bonds are made up of 1 sigma bond and 2 pi bonds.

As a rule of thumb, Pi bonds are represented by the Greek letter π in reference to the p orbitals. Observed from down the bond axis, a pi bond has the same symmetry as a p orbital. Generally, p orbitals form pi bonds, but d orbitals can too. This behavior is how we can have metal-to-metal multiple bonding (Helmenstine, 2019c).

There are some ideas and rules that you must know in order to fully understand hybridization:

1. "Hybrid orbitals do not exist in isolated atoms. They are formed only in covalently bonded atoms.

2. Hybrid orbitals have shapes and orientations that are very different from those of the atomic orbitals (s, p, d, f) in isolated atoms.

3. A set of hybrid orbitals is generated by combining atomic orbitals. The number of hybrid orbitals in a set is equal to the number of atomic orbitals that were combined (addition) to produce the set.

4. All orbitals in a set of hybrid orbitals are equivalent in shape and energy.

5. The type of hybrid orbitals formed in a bonded atom depends on its electron-pair geometry as predicted by the VSEPR theory.

6. Hybrid orbitals overlap to form σ (sigma) bonds. Unhybridized orbitals overlap to form π bonds." (OpenStax, 2016).

Molecules that contain double or triple bonds can be explained using the hybridization model. The carbon atoms in ethene (C_2H_4), for example, are linked by two covalent bonds, and the hydrogen atoms are linked to the carbon atoms by single bonds (Chemistry LibreTexts & Editors, 2016).

A sigma bond (σ bond) is a bond that is formed by the overlapping of orbitals in an "end-to-end fashion," (Chemistry LibreTexts & Editors, 2016) with the areas of the highest electron density primarily found between the nuclei of the bonding atoms. The pi bond (π bond), on the other hand, is formed between atoms when their orbitals overlap in a side-by-side fashion, with the areas of high electron density concentrating above and below the plane of their nuclei (Chemistry LibreTexts & Editors, 2016).

In C2H4, there are a total of six sigma bonds— three bonds for each carbon atom. Pi bonds extend above and below the plane of the molecule and are the second bond between the carbon atoms of a molecule. All six atoms are contained in this plane, as well as all sigma bonds.

The form of chemical bonds in diatomic molecules can be explained by thinking in terms of overlapping atomic orbitals. However, we need to use a more detailed model of molecules with more than two atoms to understand how they can form stable bonds. Let's take a look at a more complicated molecule and see why that is. Water, a polyatomic molecule, has one oxygen atom bonding to two hydrogen atoms (OpenStax, 2016).

"Oxygen has the electron configuration $1s^2$ $2s^2$ $2p^4$, and contains two unpaired electrons (one in each of the two 2p orbitals). Valence bond theory would predict that the two O–H bonds form from the overlap of these two 2p orbitals with the 1s orbitals of the hydrogen atoms. If this were the case, the bond angle would be 90° because p orbitals are perpendicular to each other. Experimental evidence shows that the bond angle is 104.5°, not 90°. The prediction of the valence bond theory model does not match the real-world observations of a water molecule; a different model is needed" (OpenStax, 2016).

In Lewis electron dot structures, double bonds are represented by double lines similar to an equal sign (=). It is important to note the differences between these lines. One represents a sigma bond, and the other represents a pi bond. In single bonds (—), only a sigma bond is represented. In triple bonds, which are depicted as 3 lines, one is a sigma bond, while the other two are pi bonds (Chemistry LibreTexts & Editors, 2016).

Metallic Bonds

The sea of electrons theory was introduced by Paul Drüde in the early 1900's by modeling metals as a mixture of atomic cores (atomic core= positive nuclei + inner electron shells) and valence electrons (Clark, 2019). Logically enough, metal atoms are the only atoms capable of forming metallic bonds. Metallic bonds join a bulk of metal atoms rather than ionic bonds that link metals to nonmetals. Metallic bonds are essentially the middle ground between ionic and covalent bonds in this regard.

The sea of electrons model is an oversimplified view of metallic bonding. There are many other factors at play including pressure levels and energy states (Helmenstine & ThoughtCo, 2020b).

Many of the characteristic properties of metals are attributable to the non-localized 'sea' of the valence electrons. For example, the free roaming nature of these electrons are responsible for the high conductivity of metals (The Editors of Encyclopaedia Britannica & L. Hosch, 2019).

Metals tend to have a strong bond between their atoms, as evidenced by their high melting points and extremely high boiling points. Even softer metals (E.g, sodium, which has a melting point of 97.8°C) melt at a far higher temperature than the element neon, for example, which has a melting point of -248.6°C.

Let's keep using sodium as our example element. When sodium atoms form a bond, "the electron in the 3s atomic orbital of one sodium atom shares space with the corresponding electron on a neighboring atom to form a molecular orbital - in much the same sort of way that a covalent bond is formed" (Clark, 2019).

The difference between the bonds arises in just how many atoms are involved in each bond. Each atom is, in a sense, crowded in by eight others that all want a share of its pie.

"...each sodium atom is being touched by eight other sodium atoms - and the sharing occurs between the central atom and the 3s orbitals on all of the eight other atoms. Each of these eight is in turn being touched by eight sodium atoms, which in turn are touched by eight atoms - and so on and so on, until you have taken in all the atoms in that lump of sodium. All of the 3s orbitals on all of the atoms overlap to give a vast number of molecular orbitals that extend over the whole piece of metal" (Clark, 2019).

In these molecular orbitals, electrons have free movement, so each one can drift apart from its parent atom. Those that do drift are called delocalized electrons (Helmenstine & ThoughtCo, 2020b). Strong forces of attraction hold the metal atoms together between their positive nuclei and their delocalized electrons, sort of like how the positively charged protons in the nucleus of an atom with +1 charge is attracted to the electrons of another atom with -1 charge in ionic bonds (Clark, 2019).

Practice Questions

1. Question: "Although element 114 is not stable enough to occur in nature, two isotopes of element 114 were created for the first time in a nuclear reactor in 1999 by a team of Russian and American scientists. The element is named after the Flerov Laboratory of Nuclear Reactions of the Joint Institute for Nuclear Research in Dubna, Russia, where the element was discovered in 1998. The name of the laboratory, in turn, honors the Russian physicist Georgy Flyorov. Write the complete electron configuration for element 114" (Chemistry LibreTexts, 2017).

Remember to use the diagram of orbital order and the periodic table for the element's atomic number.

Answer: $1s^2\ 2s^2\ 2p^6\ 3s^2\ 3p^6\ 4s^2\ 3d^{10}\ 4p^6\ 5s^2\ 4d^{10}\ 5p^6\ 6s^2\ 4f^{14}\ 5d^{10}\ 6p^6\ 7s^2\ 5f^{14}\ 6d^{10}\ 7p^2$.

2. **Question: "Write the electron configuration of mercury (Atomic No. (Z) = 80), showing all the inner orbitals.**

   Using the diagram of orbital order and the periodic table as a guide, fill the orbitals until all 80 electrons have been placed" (Chemistry LibreTexts, 2017).

   Answer: "After filling the first five rows, we still have

   $(80 - 54) = 26$ more electrons to accommodate. We

   need to fill the 6s (2 electrons), 4f (14 electrons), and

   5d (10 electrons) orbitals. The result is mercury's

   electron configuration:

   $Hg\ (80) = 1s^2\ 2s^2\ 2p^6\ 3s^2\ 3p^6\ 4s^2\ 3d^{10}\ 4p^6\ 5s^2\ 4d^{10}\ 5p^6\ 6s^2\ 4f^{14}\ 5d^{10} = [Xe]6s^2\ 4f^{14}\ 5d^{10}$

   with a filled 5d subshell, a 6s2 4f14 5d10 valence shell configuration, and a total of 80 electrons. (You should always check to be sure that the total number of electrons equals the atomic number.)" (Chemistry LibreTexts, 2017).

3. **Question: Remember Aufbau's principle: "If we follow the pattern across a period from B**

(Z=5) to Ne (Z=10) the number of electrons increases and the subshells are filled. Here we are focusing on the p subshell in which as we move towards Ne, the p subshell becomes filled" (Chemistry LibreTexts, 2019).

Answer:

B (Z=5) configuration: $1s^2\ 2s^2\ 2p^1$

C (Z=6) configuration: $1s^2\ 2s^2\ 2p^2$

N (Z=7) configuration: $1s^2\ 2s^2\ 2p^3$

O (Z=8) configuration: $1s^2\ 2s^2\ 2p^4$

F (Z=9) configuration: $1s^2\ 2s^2\ 2p^5$

Ne (Z=10) configuration: $1s^2\ 2s^2\ 2p^6$

4. **Question: Potassium iodide has what type of bond?**

Possible Answers:

Ionic bond

Polar covalent bond

Nonpolar covalent bond

Polar ionic bond

Answer: Ionic bond

Explanation:

"Potassium iodide (KI) forms an ionic bond. Potassium and iodine have very different electronegativities. The two atoms would form an ionic bond since ionic bonds form between atoms with a large difference in electronegativity, approximately a difference of >1.7 using the Pauling scale will result in an ionic bond" (Varsity Tutors, n.d.-b).

5. **Question: Out of the following definitions, which is the best description of an ionic bond?**

"Possible Answers:

An attractive force that exists between two atoms that have one or more of the same electrons in their valence shells.

Positive and negative poles that form on a molecule that is made of two nonmetals with different electronegativities.

An attractive force that exists between a cation, which has lost electrons, and an anion, which has gained electrons.

An attractive force that exists between a cation, which has gained electrons, and an anion, which has lost electrons.

Answer: An attractive force that exists between a cation, which has lost electrons, and an anion, which has gained electrons" (Varsity Tutors, n.d.-b).

Explanation:

"An ionic bond is an electrostatic force between a positively charged cation (formed when electrons are lost) and a negatively charged anion (formed when electrons are gained). While one answer choice mentions positive and negative, it is actually a description of a polar covalent bond. The easiest way to recognize this is the fact that it specifies a bond between two nonmetals. A metal is generally a component of an ionic compound" (Varsity Tutors, n.d.-b).

6. **Question: Which one of the following statements is true about ionic compounds?**

"Possible Answers:

Water is an example of an ionic compound

They are formed by molecules that equally share electrons

They are non-polar

They are 3D arrays of charged molecules" (Varsity Tutors, n.d.-b).

Answer- They are 3D arrays of charged molecules

Explanation:

"Ionic compounds are made of atoms that are held together by ionic bonds. These bonds allow charged particles that have very different electronegaticities to stick together, forming 3D crystals (lattice structures). For example, potassium and bromine form potassium bromide, an anti-convulsant. The electrons are shared very disproportionately between the positive and negatively charged particles, with all the electrons located on the more electronegative atom. Ionic compounds are not non-polar because non-polar is used to describe a molecule, whereas ionic compounds are not molecules. Water is an example of a polar molecule but is not an ionic compound since the hydrogen atoms and oxygen atom share their electrons" (Varsity Tutors, n.d.-b).

7. **Question: Which of the following characteristics does *not* describe covalent compounds?**

"Possible Answers:

They contain bonds formed by sharing one or more pairs of valence electrons between atoms.

They are poor electrical conductors.

Many are not soluble in water, but are soluble in nonpolar liquids.

The elements involved generally have large differences in electronegativity.

They have low melting and boiling points.

Answer: The elements involved generally have large differences in electronegativity" (Varsity Tutors, 2015).

"Explanation:

The incorrect statement is a property of ionic compounds rather than covalent. Recall that electronegativity is a measure of the ability of an atom to draw electrons to itself. Ionic compounds are formed by elements with very different electronegativities, since elements with different electronegativities will tend to form positive and negative ions (that is, they give away or gain electrons easily). In contrast, covalent bonds are formed by elements that are close in electronegativity and could exist as stable free molecules. All other statements are true of covalent compounds" (Varsity Tutors, 2015).

8. **Question: Complete the sentence: When covalent bonds are formed _____________.**

"Possible Answers:

Atomic species either surrender or gain electrons.

Atomic species share electrons equally.

Atomic species share electrons, but not necessarily equally.

Atomic species share their outermost electrons, allowing them to conduct electricity.

Atomic species share inner and outer shell electrons.

Answer: Atomic species share electrons, but not necessarily equally" (Varsity Tutors, 2015).

"Explanation:

Covalency is a form of electron sharing that lets an atom fulfill the octet rule. The sharing may be unequal, in which case the more electronegative species more strongly attracts the electrons than the weaker, less electronegative species, creating a polar covalent bond. In the Lewis dot model, however, we draw the electrons in freeze-frame, equally distributed between the various atoms" (Varsity Tutors, 2015).

9. **Question: "What type of bond does O_2 have?**

   Possible Answers:

   Ionic bond.

   Nonpolar covalent bond.

   Polar covalent bond.

   Polar ionic bond.

   Answer: Nonpolar covalent bond.

   Explanation:

   O_2 has a nonpolar covalent bond. Nonpolar covalent bonds are bonds formed between atoms that have the same (or nearly the same) electronegativity. Since both oxygen atoms have the same electronegativity, they will have a nonpolar covalent bond between them" (Varsity Tutors, 2015).

10. **Question: Which of the following properties determines the polarity of a covalent bond?**

    Possible Answers:

The molecular geometry

The electronegativities of the atoms involved

The electronic geometry of the compound

The atomic radii of the atoms involved

Answer: The electronegativities of the atoms involved

"Explanation:

Polarity in a bond results from an uneven sharing of electrons. An extreme example is an ionic bond, in which an electron is almost fully transferred from one atom to another due to the second atom's electron affinity.

In order to generate an uneven pull on shared electrons, the atoms involved in the bond must have significantly different electronegativities. This will cause one atom to pull electrons closer to its nucleus, away from the other atom involved.

Molecular and electronic geometry can affect the polarity of a compound, but do not directly affect the polarity of a single given bond. Atomic radius does not play a significant role in polarity" (Varsity Tutors, 2015b).

11. Question: Which of the following compounds is not held together by ionic bonds?

"Possible Answers:

Mg

F2

CO2

MgO

NaCl

Answer: CO2

Explanation:

Covalent bonds are formed when two nonmetals are bonded together. This covalent bond means that the electrons are shared by the two atoms in order to satisfy each atom's octet. There is very little difference in the electronegativities of the two atoms involved in the bond, so neither atom pulls the electrons closer to its nucleus.

Ionic bonds are formed between a metal and a nonmetal. Due to the dramatic difference between the electronegativities of metals and nonmetals, the electrons are pulled tightly to the nonmetal and away from the metal nucleus. This results in each atom having a full octet, even though the electrons are not shared.

Carbon and oxygen are both nonmetals, so we would expect only covalent bonds in carbon dioxide" (Varsity Tutors, 2015b).

12. Question: What type of bond does NaCl have?

"Possible Answers:

Ionic

Polyatomic ion

Covalent

Metallic

Answer: Ionic

Explanation:

Ionic compounds are formed between metals, which want to lose electrons, and nonmetals, which want to gain electrons. The sodium (Na) will completely transfer an electron to the chlorine (Cl), giving both atoms a complete octet without sharing electrons.

Covalent compounds generally form between two nonmetals that will both share electrons to complete their octets. Metallic compounds are built from only metals. Polyatomic ions will have a formal charge" (Varsity Tutors, 2015b).

13. **Question: Which of these bonds is the strongest?**

Possible Answers:

Double bond.

Triple bond.

Coordinate covalent bond.

Single bond.

Answer: Triple bond

"Explanation:

Triple bonds involve sharing a total of six electrons. They are the shortest in length and store the most energy, making them difficult to break. These properties make triple bonds stronger than double or single bonds.

Coordinate covalent bonds form when one atom contributes two electrons to be shared between nuclei, as opposed to each atom sharing a single electron. Once formed, coordinate covalent bonds have essentially identical properties to any other single bonds" (Varsity Tutors, 2015b).

14. **Question: Which of these molecules do you predict will have the shortest distance between their atomic nuclei?**

Possible Answers:

H_2

HCl

N_2

O_2

Answer: N_2

"Explanation:

Diatomic nitrogen, N_2, contains a triple bond between the two nitrogen atoms. Since there are six electrons being shared, the bond is stronger and the atoms are being pulled closer together. Diatomic oxygen, O_2, has a double bond, which is not as strong. Diatomic hydrogen, H2, has a single bond, which has a bond length greater than either a double or triple bond. Hydrochloric acid, HCl, is an ionic compound and will have a bond length comparable to a single covalent bond" (Varsity Tutors, 2015b).

15. **Question: How many bonds are holding this molecule together?**

CH2CHCl

Possible Answers:

Four sigma bonds and two pi bonds.

Five sigma bonds and one pi bond.

Four sigma bonds and one pi bond.

3 sigma bonds and 3 pi bonds.

Five sigma bonds and two pi bonds.

Answer: Five sigma bonds and one pi bond.

"Explanation:

The molecule has four single bonds and one double bond.

The first carbon forms two single bonds with the two hydrogen atoms and a double bond with the second carbon. The second carbon has single bonds with the third hydrogen and the chlorine atom. In total, each carbon has two single bonds and one double bond (to each other), for a total of four single bonds and one double bond.

Each single bond contains one sigma bond and zero pi bonds. Each double bond contains one sigma bond and one pi bond. This molecule has five sigma bonds and one pi bond" (Varsity Tutors, 2016).

Chapter 4: Chemical Reactions

What happens to matter when chemical changes occur? How do we represent these changes in a concise yet accurate manner while following the law of the conservation of mass? Enter chemical equations and stoichiometry.

Chemical equations are the written form of chemical reactions. They can exist in the form of word equations or symbol equations.

Word equations use the full names of elements and words for their interactions. Chemical equations use the symbols of elements and symbols (+, arrows, etc.) to show what elements are having what kind of reaction and with what results.

For example, the word equation that shows hydrogen (H) combining with chlorine (Cl) to create hydrochloric acid (HCl) is: Hydrogen gas [and] Chlorine gas [combine to make] Hydrochloric acid.

The chemical equation, which we will be using instead of word equations for the rest of this chapter, is as follows: H + Cl → HCl.

The order in which you write your reactants does not matter. However, it is important to know that not all equations you come across will be *balanced.* What does a balanced chemical equation look like? The one that we just looked at is a good example. An unbalanced version of the exact same equation might look like this: $H_2 + Cl_2 \rightarrow HCl$.

This version is unbalanced because there are differing amounts of atoms on either side of the combine-to-make symbol ($\rightarrow$). Unbalanced chemical equations are pretty much useless to us. They only have functionality once they are balanced. This is in line with the law of the conservation of mass (Helmenstine, 2020g). Having an unbalanced equation means that atoms have either been created or destroyed in the process of combining. Thus, we have to balance such equations to make them accurate and usable.

So, how do we go about balancing equations? Let's go back to the unbalanced HCl equation.

$H_2 + Cl_2 \rightarrow HCl$.

The first thing to note here is that we have 2 hydrogen atoms and 2 chlorine atoms before the reaction occurs, but only 1 of each afterward. This unbalanced nature can be changed, but not by changing the subscripted values. The subscripted values in any equation relate to the number of atoms in a molecule. Changing this will not only change the number of atoms that you have, but it will also change the type of molecule you come up with after a reaction, which we don't want.

So, instead of changing the subscripted values to balance a chemical equation, we add coefficients. These act as whole number multipliers instead of individual atoms in the way that subscripts do (Helmenstine, 2020g). So in order to balance the aforementioned equation, we do the following:

Step 1- $H_2 + Cl_2 \rightarrow HCl$ (Unbalanced, 2 H and 2 Cl atoms before reaction, 1 after).

Step 2- $H_2 + Cl_2 \rightarrow 2HCl$ (Add coefficient to product)

$H_2 + Cl_2 \rightarrow 2HCl$ is a balanced equation. This is equivalent to $2(HCl)$

Why? There are an equal number of hydrogen and chlorine atoms on either side of the arrow.

In chemical equations, the state (solid, liquid, or gas) of an element can be superscripted. These will be represented as (s) for solids, (l) for liquids, and (g) for gases. Incorporating these subscripts (which are optional unless otherwise stated) into the previously shown equation looks like this: $H_2(g) + Cl_2(g) \rightarrow 2HCl(g)$.

There are also multiple types of chemical reactions. The basic type, called combination reactions and/or synthesis reactions, are those that we have just taken a look at.

Some of the most common combination reactions are those that occur between oxygen and another element to form an oxide. Most metals and nonmetals react readily with oxygen under the majority of naturally occurring conditions. One example we can look at is magnesium oxide (MgO). Magnesium is quite a flammable element, and combining it with oxygen from the air under many conditions will produce a fine powder composed of magnesium oxide (LibreTexts & Soult Ph.D, 2019).

2Mg(s) + O2(g) → 2MgO(s).

There are still a further 5 types that we will be going through for the majority of this chapter.

Decomposition Reactions

You can think of decomposition reactions as the inverse of combination reactions. Instead of combining reactants to create a product, we are breaking down a product back into its reactants. Again, remember the law of the conservation of mass.

A decomposition reaction is a reaction that shows a compound breaking down into at least two simpler substances. Most decomposition reactions are formulated as: AB→A+B. Decomposition reactions will most often require an input from an energy source (heat, light, or electricity). Compounds that are composed of just two elements are called binary compounds. The decomposition of a binary compound into its elements is the simplest kind of decomposition reaction that you will come across (LibreTexts & Soult Ph.D, 2019).

Mercury oxide, for example, decomposes into its products: mercury and oxygen gas, when exposed to an adequate amount of heat.

2HgO(s) → 2Hg(l) + O2(g)

Some compounds can react to produce another set of compounds. Thus, a reaction that leaves some of its products as compounds is still classed as a decomposition reaction. An example of this is how "a metal carbonate decomposes into a metal oxide and carbon dioxide gas. For example, calcium carbonate decomposes into calcium oxide and carbon dioxide" (LibreTexts & Soult Ph.D, 2019).

$$CaCO_3(s) \rightarrow CaO(s) + CO_2(g)$$

Note that the number of compounds left after the decomposition reaction of some compounds is not limited to just two. Multiple compounds can be created and classified as products, just not elemental products.

Another example of this is when metal hydroxides decompose into metal oxides and water upon being heated up to an adequate level. Sodium hydroxide, for example, decomposes to produce sodium oxide and water (LibreTexts & Soult Ph.D, 2019).

$$2NaOH(s) \rightarrow Na_2O(s) + H_2O(g).$$

Displacement Reactions

Single-Displacement Reactions

Compounds undergo single-displacement reactions when one element replaces another element similar to itself. Generally, single-displacement (also called single-replacement) reactions look like or similar to the following: A + BC → AC + B

In this reaction, element A represents a metal and replaces another metal, which is represented by element B. If the element that is being replaced is a nonmetal, it must be replaced by another nonmetal. The general equation then becomes: Y + XZ → XY + Z.

Y is a nonmetal and replaces the nonmetal Z in the compound with X.

When we look at a chemical reaction in the form of an equation, we are able to tell whether or not it is a single-displacement reaction based on its character trait: "*one cation or anion trading places with another to form a new product*" (Helmenstine, 2019a). The differences between an element and a compound in an equation are relatively easy to spot based on what we learned about combination equations or reactions.

Here's an example: "Magnesium is a more reactive metal than copper. When a strip of magnesium metal is placed in an aqueous solution of copper (II) nitrate, it replaces the copper. The products of the reaction are aqueous magnesium nitrate and solid copper metal" (LibreTexts & Soult Ph.D, 2019).

Mg(s) + Cu(NO3)2(aq) → Mg(NO3)2(aq) + Cu(s)

Many metals will readily react with acids, and when they do, one of the products of the reaction is hydrogen gas. For example, when zinc reacts with hydrochloric acid, it will produce aqueous zinc chloride and hydrogen.

$$Zn(s) + 2HCl(aq) \rightarrow ZnCl_2(aq) + H_2(g).$$

What about when two compounds react? Usually, when this happens, both cations or both anions will change partners, producing a double-displacement reaction (Helmenstine, 2019a). Let's take a closer look at double-displacement.

Double-Displacement Reactions

A double-displacement reaction occurs when the positive and negative ions (cations and anions) of two ionic compounds exchange places to form two new compounds (LibreTexts & Soult Ph.D, 2019). This type of reaction most commonly occurs between ionic compounds. However, the bonds that form between the chemicals may be either ionic or covalent in nature (technically speaking), but we will be focusing on the ionic compounds. Acids or bases also participate in these reactions, but we will be taking a more in-depth look at acids and bases in a later chapter.

The bonds formed in the molecules of the reactant and the bonds formed in the product compounds are the same. The most common solvent for this type of reaction is water. The general form of a double-displacement (also called double-replacement, salt metathesis, or exchange) reaction is:

$$AB + CD \rightarrow AD + CB$$

During this reaction, A and C are the positively charged ions (or cations), whereas B and D are the negatively charged ions (or anions). It is common for substances in aqueous solution to undergo double-replacement reactions. When a reaction occurs, one or more of its products is usually a solid precipitate, a gas, or a molecule, such as water (LibreTexts & Soult Ph.D, 2019).

If the cations exchange anions with one another, you know that you are dealing with a double-displacement reaction. This is the easiest way to identify double-displacement reactions, but if you would like to double check (if the states of matter are subscripted), look out for aqueous reactants (aq) and the formation of a solid product as double-displacement reactions typically produce a precipitate (Helmenstine, 2011).

"A precipitate forms in a double-replacement reaction when the cations from one of the reactants combine with the anions from the other reactant to form an insoluble ionic compound. When aqueous solutions of potassium iodide and lead (II) nitrate are mixed, the following reaction occurs: 2KI(aq) + Pb(NO3)2(aq) → 2KNO3(aq) + PbI2(s)" (LibreTexts & Soult Ph.D, 2019).

Combustion Reactions

We can identify a combustion reaction by whether or not one of the reactants is oxygen gas, releasing energy in the form of light and heat. All combustion reactions have O2 as one of the reactants. The combustion of hydrogen gas, for example, will produce water vapor (LibreTexts & Soult Ph.D, 2019). We don't need to specify that hydrogen will combust with oxygen, as that is implicit with the use of the word "combustion".

$$2H_2(g) + O_2(g) \rightarrow 2H_2O(g)$$

Another way of looking at combustion is as a chemical reaction that takes place "between a fuel and an oxidizing agent that produces energy, usually in the form of heat and light. Combustion is considered an exergonic or exothermic chemical reaction. It is also known as burning" (Helmenstine & ThoughtCo, 2020a).

Exergonic means that it involves the release of energy, and exothermic means that "the reaction may occur spontaneously and result in higher randomness or entropy ($\Delta S > 0$) of the system. They are denoted by a negative heat flow (heat is lost to the surroundings) and a decrease in enthalpy ($\Delta H < 0$). In the lab, exothermic reactions produce heat or may even be explosive" (Helmenstine & ThoughtCo, 2019). Combustion often involves an intense release of heat and energy in the form of an explosion, often occurs spontaneously, and has a negative heat flow due to a loss of energy (in the form of heat) to the surrounding environment.

Due to the double bond between the oxygen atoms in O_2 being weaker than single bonds as well as other double bonds, combustion releases heat (Helmenstine & ThoughtCo, 2020a). While the energy produced is absorbed during the reaction, it is released later when stronger bonds are formed to produce carbon dioxide (CO_2) and water (H_2O). While the fuel does have a role in the production of energy during the reaction, it is minor in comparison to that produced by the forming of stronger bonds because the chemical bonds in the fuel are similar to the level of energy in the bonds in the products (Helmenstine & ThoughtCo, 2020a).

Many combustion reactions will involve a hydrocarbon (a compound that is made up of carbon and hydrogen), but this reactant is not necessary to the identification of combustion reactions in the way that oxygen gas is. When a hydrocarbon is involved in combustion, we know that its products will always include carbon dioxide and water (LibreTexts & Soult Ph.D, 2019).

There is an abundance of hydrocarbons that you have likely encountered and used in the form of fuel due to their combustion releasing a very large amount of heat energy. For example, propane (C_3H_8) is a gaseous hydrocarbon that is commonly used as the fuel source in gas grills.

$$C_3H_8(g) + 5O_2(g) \rightarrow 3CO_2(g) + 4H_2O(g)$$

Redox Reactions

If you have ever walked across a rusty bench or picked up a greening copper coin, you have seen the end result of a redox reaction. Redox, as you may know, is simply a shortened version of the name "reduction oxidation".

Simply put, oxidation involves the loss of electrons from an atom, while reduction involves the gain of electrons by an atom. This may be confusing at first, as when we talk about the reduction of something, we tend to be referring to a loss, not a gain. But when it comes to redox reactions, the reduction aspect refers to a loss in *charge,* not in electrons. Gaining an electron, due to their negative charge, causes a negative net charge in the atom that gains it.

In a redox reaction, this gain-and-loss occurs at the same time, so that one element or compound gains electrons while the other loses electrons (Ball & Key, 2014). While we can separate redox reactions into these two separate occurrences, it is important to remember that they always take place at the same time. It is only through abstraction via fantasy that we are able to separate them and make the whole process easier to understand.

When working with redox reactions, it is vital that we keep track of the electrons assigned to each atom during the reaction. How? By once again using our imagination. The number of electrons in an atom is counted by an artificial count called the oxidation number (Ball & Key, 2014). Various rules determine the oxidation number of an atom. Something else to bear in mind is that oxidation numbers have little to no correlation or causation with the charge of the atom. Do not confuse the concepts of charge and oxidation numbers; doing so could cost you dearly in the future.

As mentioned, there are rules that govern the assignment of oxidation numbers to atoms. They are as follows:

1. "Atoms in their elemental state are assigned an oxidation number of 0.

2. Atoms in monatomic (i.e., one-atom) ions are assigned an oxidation number equal to their charge. Oxidation numbers are usually written with the sign

first, then the magnitude, which differentiates them from charges.

3. In compounds, fluorine is assigned a −1 oxidation number; oxygen is usually assigned a −2 oxidation number (except in peroxide compounds [where it is −1] and in binary compounds with fluorine [where it is positive]); and hydrogen is usually assigned a +1 oxidation number (except when it exists as the hydride ion, H−, in which case rule 2 prevails).

4. In compounds, all other atoms are assigned an oxidation number so that the sum of the oxidation numbers on all the atoms in the species equals the charge on the species (which is zero if the species is neutral)." (Ball & Key, 2014).

When dealing with redox equations, bear in mind that we cannot write the loss of an electron (and thus a +1 charge) using a subtraction sign. Note: e^- = 1 electron.

This is wrong: $Na - e^- \rightarrow Na^+$

Instead, we write this equation as: $Na \rightarrow Na^+ + e^-$

A useful way of remembering this is simply to bear in mind that when an electron (e^-) is on (what is typically considered) the product side of the arrow (Reactants $\rightarrow$ *Product*), the electron is being *lost*. When it is on the reactant side, it is being gained (DeWitt, 2015a).

Redox reactions will always occur with a simultaneous change in the oxidation numbers of some if not all of the involved atoms, always in the pattern of one increasing and the other decreasing. Hence, in order for a redox reaction to occur, at least two elements must change their oxidation numbers (Ball & Key, 2014). This has to be the case for both oxidation and reduction to occur at once, and it will help you identify redox reactions.

A molecule is oxidized when its oxidation number increases during a redox reaction, and when the oxidation number of an atom decreases during the course of a redox reaction, that atom is being reduced. Based on this, we are able to define oxidation and reduction in terms of increasing or decreasing oxidation numbers, respectively (Ball & Key, 2014).

Here are some examples for you to go through, taken from Ball & Key on OpenTextBC:

"In H_2, both hydrogen atoms have an oxidation number of 0, by rule 1.

In NaCl, sodium has an oxidation number of +1, while chlorine has an oxidation number of −1, by rule 2.

In H_2O, the hydrogen atoms each have an oxidation number of +1, while the oxygen has an oxidation number of −2, even though hydrogen and oxygen do not exist as ions in this compound as per rule 3.

By contrast, by rule 3 in hydrogen peroxide (H2O2), each hydrogen atom has an oxidation number of +1, while each oxygen atom has an oxidation number of −1.

We can use rule 4 to determine oxidation numbers for the atoms in SO2.

Each oxygen atom has an oxidation number of −2; for the sum of the oxidation numbers to equal the charge on the species (which is zero), the sulfur atom is assigned an oxidation number of +4.

Does this mean that the sulfur atom has a 4+ charge on it? No, it only means that the sulfur atom is assigned a +4 oxidation number by our rules of apportioning electrons among the atoms in a compound" (Ball & Key, 2014).

Collision Theory

Collision theory states that "the rate of a chemical reaction is proportional to the number of collisions between reactant molecules" (Harper College, 2019).

Thus, we use collision theory to predict the rates of chemical reactions, particularly those that take place between gases. For collision theory to apply, we assume that the reacting atoms or molecules have collided in order for a reaction to take place (Augustyn & Encyclopaedia Britannica, 2021).

Most collisions, however, do not produce chemical change. It is only possible to produce chemical change in a collision if the species brought together possess a sufficient amount of internal energy, equal to the activation energy of the reaction. This need for sufficient energy levels causes a change in temperature to affect the reaction rate. As we know, heat is directly proportional to energy, and thus higher temperatures mean more effective molecular collisions. In addition, the colliding species must be oriented in a way that allows atoms and electrons to rearrange as needed. If molecules collide more often, the reaction will occur more often, and the reaction rate will thus increase. Effective collisions are mere fractions of the total collisions, and we classify them as those that lead to a chemical change (Harper College, 2019). Applied collision theory can only determine atomic or molecular frequencies for gases (by using kinetic theory), which means the theory is limited to gas-phase reactions (Augustyn & Encyclopaedia Britannica, 2021).

Every second, about 10^{33} collisions occur in one cubic centimeter of space between gases at room temperature and normal atmospheric pressure. Now imagine if the collision to reaction rate was 100% and every collision yielded products. All reactions would be completed in the blink of an eye. Unsuccessful collisions can be likened to a pair of billiard balls colliding. Nothing really happens; they merely bump against each other and go on their way. If a reaction does not occur between A and B during a collision, it is because the reaction requires a significant disruption or rearrangement of the bonds between their atoms that has not been achieved during the collision. "In order to effectively initiate a reaction, collisions must be sufficiently energetic (or have sufficient kinetic energy) to bring about this bond disruption" (Lawson et al., 2020).

Consider the following bimolecular equation:

A + B → Products

As molecules A and B approach each other to react, some of their bonds will be disrupted and others will be created for the forming of new products. Such a successful encounter is called a collision (Lawson et al., 2020).

There is a proportional relationship between the frequency of collisions between reactants A and B in a gas and its concentration; if reactant [A] doubles, the frequency of A-B collisions will double; the same goes if you were to double the amount of reactant [B] (Lawson et al., 2020). "If all collisions lead to products, then the rate of a bimolecular process is first-order in A and in B, or second-order overall: rate = k[A][B]" (Lawson et al., 2020).

Practice Questions

1. Question: Balance this chemical equation:

$Pb(NO_3)_2(aq) + KCl(aq) \rightarrow PbCl_2(s) + KNO_3(aq)$

Possible Answers:

2Pb(NO3)2(aq) + 4KCl(aq) → PbCl2(s) + 4KNO3(aq)

Pb(NO3)2(aq) + 2KCl(aq) → PbCl2(s) + 2KNO3(aq)

Pb(NO3)2(aq) + 2KCl(aq) → PbCl2(s) + KNO3(aq)

Pb(NO3)2(aq) + KCl(aq) → PbCl2(s) + 2KNO3(aq)

Answer: Pb(NO3)2(aq) + 2KCl(aq) → PbCl2(s) + 2KNO3(aq)

"Explanation:

A balanced chemical equation will have the same number of each atom on both sides of the reaction.

Pb(NO3)2(aq)+KCl(aq)→PbCl2(s)+KNO3(aq)

Start by balancing the number of nitrate molecules (NO3). Since we have 2 nitrate on the left, we must also have 2 nitrate on the right.

Pb(NO3)2(aq)+KCl(aq)→PbCl2(s)+2KNO3(aq)

This equation now gives 1 potassium (K) on the left, and 2 potassium on the right. Balance the potassium.

Pb(NO3)2(aq)+2KCl(aq)→PbCl2(s)+2KNO3(aq)

This equation is balanced because there are equal numbers of each atom on both sides" (Varsity Tutors, 2012).

2. Question: "When an electric current is passed through pure water, it decomposes into its elements. Write a balanced equation for the decomposition of water" (Chemistry LibreTexts, 2022).

Answer:

"Step 1: Plan the problem.

Water is a binary compound composed of hydrogen and oxygen. The hydrogen and oxygen gases produced in the reaction are both diatomic molecules.

Step 2: Solve.

The skeleton (unbalanced) equation:

$H2O(l) \rightarrow H2(g) + O2(g)$

(In some nomenclature, the abbreviation "elec" would appear above the arrow to indicate the passage of an electric current to initiate the reaction.)

Balance the equation.

$2H2O(l) \rightarrow 2H2(g) + O2(g)$

Step 3: Think about your result.

The products are elements and the equation is balanced" (Chemistry LibreTexts, 2022).

3. Question: (This question also contains additional learning— the link between redox reactions and single displacement) "Which of the following reactions involves both reduction and oxidation reactions?" (Chemistry Varsity Tutors, n.d.).

"Possible Answers:

The reaction of sodium chloride and calcium sulfate

Single displacement reactions

Neither single displacement reactions nor the reaction of sodium chloride and calcium sulfate

Both single displacement reactions and the reaction of sodium chloride and calcium sulfate

Answer: Single Displacement Reactions.

Explanation:

A reaction that has both reduction and oxidation half reaction is called a redox reaction. It involves one or more atoms gaining electrons (reduction) and one or more atoms losing electrons (oxidation). Recall that single displacement reactions involve the replacement of an element in a compound with another element. An example of a single replacement reaction is shown below:

Na + CaSO4 → Na2SO4 + Ca

In this reaction, a sodium atom replaces a calcium atom in calcium sulfate. If we calculate the oxidation numbers for sodium and calcium, we can see that sodium loses an electron (the oxidation state goes from 0 to +1) whereas the calcium ion gains two electrons (goes from +2 to 0); therefore, this is a redox reaction. All single displacement reactions follow this general trend and are characterized as redox reactions.

The reaction equation of sodium chloride and calcium sulfate is as follows:

NaCl + CaSO4 → Na2SO4 + CaCl2

If we calculate the oxidation state of each atom, we will notice that the oxidation number doesn't change; therefore, this isn't a redox reaction" (Chemistry Varsity Tutors, n.d.).

4. Question: Predict the products for the following double-displacement reaction: AB+CD→?

Possible Answers:

AB+CD→EF+GH

AB+CD→CD+BA

AB+CD→A+B+C+D

AB+CD→AD+CB

AB+CD→AC+BD

Answer: AB+CD→AD+CB

"Explanation:

In a double displacement reaction, or double replacement reaction, opposite ions combine. What this means is that the cation of one compound will recombine and bond to the anion of the other compound and vice versa.

In this example, we should note the ideal convention of cations being written before the anion in the compound. So in AB, we can presume A is the cation and B is the anion. The same goes for CD: C is the cation and D is the anion.

With this in mind, we can now easily see how the replacement would be possible.

If A is a cation, and was originally bound to B (anion), the only other anion it is left with to bind is D. If C is a cation, and was originally bound to D (anion), it is only left with the option of binding to the now-free B anion" (Varsity Tutors High School Chemistry, n.d.).

5. Question & Additional Information: "Oxidation number rules: Elements have an oxidation number of 0 Group (vertical column in the periodic table) I and II – In addition to the elemental oxidation state of 0, Group I has an oxidation state of +1 and Group II has an oxidation state of +2." (Butane.Chem, n.d.).

"Hydrogen –usually +1, except when bonded to Group I or Group II, when it forms hydrides, -1.

Oxygen – usually -2, except when it forms a O-O single bond, or a peroxide, when it is -1.

Fluorine is always -1. Other halogens are usually -1, except when bonded to O.

1. Assign oxidation numbers to each of the atoms in the following compounds:

Na2CrO4: Na = +1 O = -2 Cr = +6

K2Cr2O7: K = +1 O = -2 Cr = +6 CO2 O = -2 C = +4
CH4 H = +1 C = -4

HClO4: O = -2 H = +1 Cl = +7

MnO2: O= -2 Mn= +4

SO_3^{2-}: O= -2 S= +4

SF4: F= -1 S= +4

a. What is the range of oxidation states for carbon?
Answer: -4 to +4

b. Which compound has C in a +4 state? Answer:
CO2

c. Which compound has C in a -4 state? Answer:
CH4" (Butane.Chem, n.d.).

6. Question: Use collision theory to explain why a higher temperature will result in a faster reaction.

"Possible Answers:

It increases the pressure of the reactants.

It increases the surface area for reactions.

It lowers the activation energy for a reaction.

It increases the number of effective molecular collisions.

It helps orient reactants into the correct position to react.

Answer: It increases the number of effective molecular collisions" (AP Chemistry Varsity Tutors, n.d.).

"Explanation:

The temperature of a system can be manipulated to increase the reaction rate. The temperature affects the rate of the reaction because as temperature increases, so does the average kinetic energy. This means that molecules (reactants) are moving around more quickly, resulting in more frequent molecular collisions, resulting in an increased rate of product formation" (AP Chemistry Varsity Tutors, n.d.).

7. Question: "Suppose that the only elements a compound contains are carbon, hydrogen, and oxygen. If complete combustion of 70.86g of this compound produces 103.9g of CO_2 and 42.52g of H_2O, what is this compound's molecular formula?

Note: The molecular weight of this compound is 120.104 g/mol." (College Chemistry Varsity Tutors, n.d.).

"Possible Answers:

$C_7H_4O_2$

$C_5H_{12}O_3$

$C_3H_4O_5$

$C_4H_8O_4$

Answer: $C_4H_8O_4$

Explanation:

To answer this question, it's important to remember the details of combustion reactions. In the presence of oxygen and a sufficient amount of energy to initiate the process, combustion reactions that go to completion will result in the complete oxidation and breakdown of the initial reactant.

The CO_2 produced from the reaction contains all of the carbon in the original reactant. Likewise, the H_2O produced contains all the hydrogen in the original compound. The oxygen content in the original compound is everything left over.

The first step, then, is to use the mass of each product to find the mass of each element in the original compound. To do this, we'll need to use the molar mass for each of the products as well as the individual elements:

103.9g CO_2 * 12 $^g/_{mol}$ C ÷ 44 $^g/_{mol}$ CO_2= 28.34g C (Simplified: 103.9(12 ÷ 44)= 28.34)

42.52g H_2O * 2 $^g/_{mol}$ H ÷ 18 $^g/_{mol}$ H_2O= 4.72g H

The combined mass of carbon and hydrogen in the original compound is:

28.34g + 4.72g= 33.06g.

The remaining mass will come from oxygen. Thus, the original compound will contain 70.86g – 33.06g= 37.8g of oxygen.

Once we have the mass of each element, we can use that information to calculate the number of moles of each element in the original compound.

$28.34g \ C \ * \ 1 \ mol \ C/12g \ C = 2.36$ mol C (Simplified: $28.34/1212 = 2.36$)

$4.72g \ H \ * \ 1 \ mol \ H/1g \ H = 4.72$ mol H

$37.8g \ O \ * \ 1 \ mol \ O/16g \ O = 2.36$ mol O

Now that we have the relative molar amounts of each element in the original compound, we can find the empirical formula by finding the smallest whole number ratios. We divide each subscript by the smallest value found (which, in this instance, is 2.36) and thus we have our empirical formula.

$C_{2.36}H_{4.72}O_{2.36} = CH2O$

From this, we can conclude that the empirical mass of the compound is 30g by multiplying each element by its mass and adding them together.

C(12) + H(2*1, as there are two hydrogen atoms) + O(16)= 30g.

Finally, since we are given the compound's molecular weight, we can use that information to find the molecular formula for the compound.

$120.104 \div 30 \approx 4$

$4(CH_2O) = C_4H_8O_4$" (College Chemistry Varsity Tutors, n.d.).

8. Question: Given the following equation, identify the reducing agent.

$Mn^{2+}(aq) + NaBiO3(s) \rightarrow Bi^{3+}(aq) + MnO4^-(aq) + Na^+(aq)$

"Possible Answers:

MnO_4^-

$NaBiO_3$

Mn^{2+}

Bi^{3+}

Answer: Mn2+

Explanation:

Recall that a reducing agent is being oxidized and is losing electrons. Thus, when looking at the equation, look for which oxidation state is becoming more positive.

Start by assigning oxidation states to the elements.

For the reactants:

Mn^{2+} has an oxidation state of +2.

Na has an oxidation state of +1.

Bi has an oxidation state of +5

O has an oxidation state of −2.

Now, assign the oxidation state of the products. Mn has an oxidation state of +7.

Na has an oxidation state of +1.

Bi has an oxidation state of +3.

O has an oxidation state of −2.

Only Mn^{2+} undergoes a loss of electrons. It must be the reducing agent" (College Chemistry Varsity Tutors, n.d.-b).

9. Question: Identify which of the following is a double displacement reaction.

Possible Answers:

(A) $CH_4 + 2O_2 \rightarrow CO_2 + 2H_2O$

(B) $NaCl + Br^- \rightarrow NaBr + Cl^-$

(C) $Zn + Cu^{2+} \rightarrow Zn^{2+} + Cu$

(D) $CuSO_4 + SrCl_2 \rightarrow CuCl_2 + SrSO_4$

Step 1: Use the balanced chemical equation to determine if there are two products and two reactants.

(A) There are two products and two reactants. This reaction passes step 1.

(B) There are two products and two reactants. This reaction passes step 1.

(C) There are two products and two reactants. This reaction passes step 1.

(D) There are two products and two reactants. This reaction passes step 1.

Step 2: Determine if all compounds involved are composed of an anionic component and a cationic component.

(A) CH_4 is composed of two different elements, but it is not an ionic salt. Furthermore, O_2 is an elemental compound and does not have a formula like AX or BY. This is not a double displacement reaction.

(B) NaCl is a salt that can be separated into the components Na^+ and Cl^-. However, Br^- is not composed of two ions. This is not a double displacement reaction.

(C) Zn is not composed of two ions. Neither is Cu^{2+}. This is not a double displacement reaction.

(D) $CuSO_4$ can be broken into the cation, Cu^{2+}, and the anion, SO^{2-}_4. $SrCl_2$ can be broken into the cation, Sr^{2+}, and two Cl^-Cl^- anions. This might be a double displacement reaction.

Step 3: Check your work: Identify the same number and charge of ions on both reactant and product sides, as well as that the anions and cations flipped between compounds.

We eliminated options A, B, and C, but to confirm D, we should still check the charges and amounts of the ions involved. From our work in Step 2, we know the numbers and charges of the reactants.

Looking towards the products, $CuCl_2$ can be broken into the cation Cu^{2+} and two Cl^- Cl^- anions. The other product, $SrSO_4$, can be broken into the cation Sr^{2+} and the anion SO_4^{2-}.

SO_4^{2-}. There are an equal number of the same charged ions on each side, and each ion ended with a different counterion.

The correct answer is (D) $CuSO_4 + SrCl_2 \rightarrow CuCl_2 + SrSO_4$ (Chemistry Study.com, 2022).

Chapter 5: Thermodynamics and Electrochemistry

"Electrochemistry is the study of chemical processes that cause electrons to move. This movement of electrons is called electricity, which can be generated by the movement of electrons from one element to another in a reaction known as an oxidation-reduction ("redox") reaction" (Chieh & Chemistry LibreTexts, 2015).

It may sound straightforward to the majority of people. Everyone thinks they know what electricity is. But as we have observed thus far, nothing is ever as it seems.

We have already taken a look at redox reactions, but since we will be using them throughout this chapter, let's have a brief recap. Redox reactions, or oxidation and reduction reactions, involve the transfer of electrons. When properly set up, these reactions generate the power in a battery, or galvanic cell. A galvanic cell consists of at least two half cells, each of which consists of an electrode and an electrolyte solution.

A redox reaction can be divided into two halves—an oxidation reaction and a reduction reaction. Just remember that these two always occur at the same time.

Electrolysis

Electrolysis is the process by which electric current is passed through a substance to produce a nonspontaneous chemical change. To produce this electric current, half cells can be set up as simple batteries by using each half reaction (oxidation and reduction) and putting two half cells together. In some respects, galvanic cells can be considered batteries, but batteries typically contain several galvanic cells connected in parallel to provide higher voltages than a single galvanic cell (Chieh & Chemistry LibreTexts, 2015). The electrolytic cell, which is used for electrolysis, can be made from two half cells. The electric energy is used in this case to induce nonspontaneous chemical reactions, or in other words, reactions that cannot occur on their own (The Editors of Encyclopaedia Britannica, 1998b).

"Oxidation of a species (atom, ion, or molecule) is a loss of electron(s), and reduction of a species a gain of electron(s). Some conventions are used to define the Oxidation States, and oxidation of an atom causes an increase in its oxidation state. Conversely, reduction causes a decrease in its oxidation state" (Chieh & Chemistry LibreTexts, 2015).

You can now see why viewing redox reactions as two separate parts of the same equation is so helpful. Without that dichotomy, understanding electrochemistry would be even more complex.

Electrons struggle through loads using the force generated by chemical reactions that occur within batteries or galvanic cells.

Chemical energy is converted into electric energy in this way. Solutions or molten salts (liquid) can be used to carry out electrolysis as the atoms that make up molten salt, for example, are disconnected and moving about rapidly. A fluid medium is required because atoms and ions need to move. The products can be in a solid, liquid, or gas form, as are the reactants of a galvanic cell (Chieh, 2015).

"Electrodes are electric conductors that are usually made up of metal and are used as either of the two terminals of an electrically conducting medium; it conducts current into and out of the medium, which may be an electrolytic solution as in a storage battery, or a solid, gas, or vacuum... The electrode from which electrons emerge is called the cathode and is designated as negative; the electrode that receives electrons is called the anode and is designated as positive." (The Editors of Encyclopaedia Britannica, 1998b). A galvanic cell, also called a voltaic cell, provide the electric energy or current through spontaneous reactions that occur within it.

Let's take a look at an example of how this process works—the electrolysis of sodium chloride.

$NaCl \rightarrow Na + Cl_2$ (This is an unbalanced equation. Remember, chlorine is one of the diatomic molecules and can't be found alone). The equation we have just looked at is an example of a redox reaction. As electrons are moving around, they create a change (DeWitt, 2015b). How are they moving? Remember from the previous chapter that it is important for us to keep track of electron movements, and we use oxidation numbers to represent these moves and changes.

The first thing we do here is assign each element its oxidation number. We can find these numbers using the periodic table—elements in the first group (vertical column) have an oxidation number of +1, and those in the second group have an oxidation number of +2, the third group has +3, etc. Halogens (which is what chlorine is) have an oxidation number of -1 unless paired with oxygen, which will give them a +1 oxidation number. Unpaired elements in an equation have an oxidation number of 0. With this information, we can now look at the equation as:

$Na^{+1}Cl^{-1} \rightarrow Na + Cl_2$

Na^{+1} has reduced, as it has *gained* an electron and lost its *positive* oxidation number.

Cl^{-1} has oxidized, as it has *lost* an electron and lost its *negative* oxidation number.

You could also think of this as the charge of the atom, which we covered earlier. NaCl, or table salt, will not go through this process on its own, which is why we refer to these types of reactions as nonspontaneous. They require electricity, for example, for the reaction to occur. One reason for this is that Na^{+1} and Cl^{-1} are unreactive in NaCl. Na doesn't want to gain an electron, and Cl doesn't want to lose one. Thus, we can must an electrical current (or electrical energy) to force this reaction to occur.

In a battery, the negatively charged side pulls electrons in and the positively charged side pushes them out. If we were to attach an electrode to the negative side of a battery, it would become a cathode. If we attach another to the positive side of the battery, it would become an anode. Once we have our battery hooked up to a pair of electrodes, one to the positive and the other to the negative, we can then place both the anode (positive side of the battery) and the cathode (negative side of the battery) inside of a container with liquid NaCl, or molten salt. Because molten salt is quite hot (around 800°C), the electrodes you would use would have to be extremely heat resistant, otherwise they would simply melt as well. Fortunately, most metals (electrodes) have very high melting points, so we would likely not have to worry too much about our electrodes melting.

Once we have both a cathode and an anode in the molten NaCl, the electrons swarming around in the molten salt will move through the anode (where oxidation occurs), into the battery, and then back out of the battery into the cathode (where reduction occurs).

Remember—anodes pull electrons in, cathodes push them back out.

So, where do these electrons get pulled from and pushed into? In our case, in molten NaCl, which has free-moving Na and Cl atoms, the electrons are pulled from the negatively charged Cl atoms, meaning they lose their charge and become neutral. The electron is then moved through the anode and into the positive side of the battery, then out the *negative* side of the battery and into the cathode. The cathode then distributes this electron, forcing it onto a nearby positively charged Na atom (which is drawn to the cathode), and then, after gaining an electron, the Na atom loses its positive charge.

Remember how we said that Cl is a diatomic molecule? Upon losing their negative charges, which were repelling them before, a pair of Cl atoms will bond and become Cl_2 (or chlorine gas) and simply float away.

These reactions can be represented by the following chemical equations:

$$Na^+ + e^- \rightarrow Na \text{ (Reduction)}$$

$$2Cl^- \rightarrow Cl_2 + 2e^- \text{ (Oxidation)}$$

These are referred to as *half reactions*. Why? Because they are only half of the reaction, funnily enough. To get the full reaction, we must then combine these two reactions. How do we do this? Well, first, to combine these reactions, we must ensure that both equations contain the same number of electrons, which they currently do not. The oxidation reaction contains 1 electron (e^-), and the reduction reaction contains 2 electrons ($2e^-$).

We can solve this problem by simply multiplying one side of the equation by 2:

$$2(Na^+ + e^- \rightarrow Na) = 2Na^+ + 2e^- \rightarrow 2Na$$

$$2Cl^- \rightarrow Cl_2 + 2e^-$$

As we can now see, both equations contain 2 electrons ($2e^-$). To combine them, we simply place the reactants together on one side of the equation, and the products together on the other side:

$2Na^+ + 2Cl^- + 2e^- \rightarrow 2Na + Cl2 + 2e^-$

We can now cancel out electrons that occur on both sides of the equation, which is why we needed both of our half reactions to have the same number of electrons going into this combination. Cancelling out the electrons and then combining the reactants will look like this:

$2Na^+ + 2Cl^- \rightarrow 2Na + Cl2$ (Remember that negatives and positives attract).

$2NaCl \rightarrow 2Na + Cl2$ (Combine the reactants, add matter state parenthesis if needed/wanted).

$2NaCl(l) \rightarrow 2Na(l) + Cl2(g)$ (Final reaction).

Thermodynamics

Now let's take a look at thermodynamics. This is one of the areas in which chemistry and physics overlap. Thermodynamics is the study of forms of energy, whether they be thermal, electrical, chemical, or mechanical. While thermodynamics has various branches, the kind that most interests chemists is the study of how energy changes during a chemical reaction (Harvey & Chemistry LibreTexts, 2019).

The Zeroeth Law of Thermodynamics

Ah... you thought we were going to start with the *first* law. And to be fair, even the greatest chemists did until the 20th century (around 1935) when a British physicist by the name of Ralph H. Fowler first coined the term "zeroeth law." He called it this because he believed that it was more fundamental even than the other laws. This isn't to say he discovered it. It was just that the zeroeth law was almost taken for granted until he explicitly stated it and gave it a name and a place with the other laws (Helmenstine, 2019b).

The zeroeth law of thermodynamics states that "If two bodies are in thermal equilibrium with a third body, they are in thermal equilibrium with each other as well" (PSIBERG Team, 2022).

A way of expressing this can be shown as follows:

A= B= C= A.

As we have mentioned, the zeroth law is an important law in thermodynamics. It is the most fundamental. Even though it was postulated after the rest of the thermodynamic laws (1st, 2nd, and 3rd), the others had reached stages too far ahead of this new law. This law was (and arguably still is) the most basic, and thus could not be named as the fourth. That is why it was named the zeroeth law of thermodynamics (PSIBERG Team, 2022).

The First Law of Thermodynamics

Newton's first law of thermodynamics. Everyone knows it well: Energy can neither be created nor destroyed; energy can only be transferred or changed from one form to another.

When we are dealing with the first law of thermodynamics, we are interested in how energy changes. To measure and calculate this, we express the change in internal energy as the following equation:

$\Delta U = Q + W$.

ΔU is the change of the internal energy of a system (For example, a human being), Q being the the net heat transferred into the system, or "the sum of all heat transfer into and out of the system" (For example, the internal temperature maintenance of mammals) and W being the net work done by the system—that is, W is the sum of all work done *on or by* the system, or activities that the object/creature is engaged in (Urone et al., 2012).

Here are some other symbols that pertain to this equation:

"U_1 (or U_i) = initial internal energy at the start of the process

U_2 (or U_f) = final internal energy at the end of the process

Δ (delta) U = U_2 - U_1 = Change in internal energy (used in cases where the specifics of beginning and ending internal energies are irrelevant)

Q = heat transferred into (Q > 0) or out of (Q < 0) the system

W = work performed by the system (W > 0) or on the system (W < 0)" (Helmenstine, 2019b).

The equation is different in physics: $\Delta U = Q - W$.

This difference comes from the fact that chemists and physicists view this problem from different angles. The chemist views the equation from the point of the system, whereas the physicist views it from the environment that acts on a system. Don't get the two confused!

Positive Q adds energy to the system, while negative W removes energy from the system. If the amount of heat (+Q) added is greater than the amount of work done (-W) the difference is stored as internal energy. Heat transfer (Q) and Work (W) are the means by which most (if not all) systems take in and expend energy. Heat transfer, being a slightly more chaotic process, is determined by temperature differences. Work, on the other hand, is a more orderly process that involves a macroscopic force exerted through a distance (Urone et al., 2012).

Heat and work can sometimes have a similar effect, and once an increase in temperature occurs, it is impossible to tell which caused it: heat or work. Both are energy in transit, so neither heat nor work can be stored within a system.

However, both can change the internal energy of a system (U), as internal energy is completely different from Q and W. There are two ways of viewing the internal energy of a system: the molecular/atomic way and the macroscopic way. We'll be focusing on the atomic/molecular.

Simply put, internal energy is the sum of atomic and molecular mechanical energy. We have to deal with averages and distributions when observing and measuring the mechanical energy of a system, as trying to manage all the molecules and atoms is currently impossible, and in a future where it were possible, odiously boring (Urone et al., 2012).

Let's take a look at a simple example of how this law works.

Let's say you have a system of some kind, and it contains 0 joules of energy. For this system to increase its internal energy, that energy has to come from somewhere. It cannot be created, so where does it come from? The energy comes from outside the system, or the environment surrounding said system. If 100 joules of energy were to enter your system, how many would have to be provided by the environment? 100, exactly. Energy has to transfer from one place to another and will often change states (The Organic Chemistry Tutor, 2017).

This change can be expressed as:

$$\Delta U = +100J$$

There are multiple types of systems that react to the environment differently. These are open systems, closed systems, and isolated systems.

Open systems are open to both heat (energy) and matter. For example, O_2 and heat can both freely flow into an open system.

Closed systems, on the other hand, can only take in heat or energy and are closed off to matter.

Isolated systems are no-go-zones, in a sense. Energy *and* matter cannot enter or leave an isolated system. Its mass and energy levels are both fixed, constant, or unchanging.

Going back to our equation for the change in internal energy:

$\Delta U = Q + W$

When W is negative, work is being done *by* the system, and it is thus losing energy. For example, if you go for a run, you are expending energy as you burn calories to perform the action of running. This is an example of -W.

When W is positive, work is being done *on* the system and it will thus *gain* energy.

Q (or heat) works similarly in principle. When Q is positive, it means that the system (the reactants and products within it) is absorbing heat. When heat energy is absorbed, it is referred to as an endothermic process.

When Q is negative, it means that the system is losing heat energy. This is called an exothermic process. To remember these easily, just think of exothermic as expelling heat. Exo; expell. Endo; absorb.

Let's take a look at a problem now, and use the first law of thermodynamics to solve it.

You have a closed system that has 60J of work done *on* it, and thereafter *loses* 150J of heat energy.

$$\Delta U = Q + W$$

$$\Delta U = -150J + 60J$$

$$\Delta U = -90J$$

Since we lost 150J of heat, we substitute Q with -150J, and since there was 60J of work done *on* the system, we substitute the W with 60J. Due to the loss of heat energy, the *temperature* of your system will have decreased as well.

The Second Law of Thermodynamics

All things decay. All natural or manmade structures rust and crumble. All young people grow old and die. All systems, organic or not, slowly lose their energy and fade away into nothingness. If you leave an ice cube at room temperature, it will slowly turn to water. Clean your room today, and you can be rest assured that there will be more to clean by the time tomorrow arrives. We refer to this decay, or natural proclivity towards disorder, as entropy.

This second law states that nature itself prevents us from reaching certain kinds of outcomes without putting work into it, referring to nonspontaneous reactions and effects. Similarly to the first law of thermodynamics, it is closely tied to the concept of the conservation of energy (Helmenstine, 2019b).

"Certain things happen in one direction and not the other; this is called the "arrow of time" and it encompasses every area of science. The thermodynamic arrow of time (entropy) is the measurement of disorder within a system. Denoted as ΔS, the change of entropy suggests that time itself is asymmetric with respect to order of an isolated system, meaning that a system will become more disordered, as time increases" (Malley et al., 2019).

Before continuing, let us take a look at nonspontaneous and spontaneous actions so that we can distinguish the two definitively, as well as an example of heat loss during the transfer of heat between locations.

1. A ball rolling downhill.

This is spontaneous, as it can happen on its own. Gravity can naturally cause a ball to roll downhill without any external energy or exertion, such as someone kicking the ball.

2. An iron rod rusting in the presence of air and water molecules.

This is also spontaneous, as the reaction that causes rusting occurs with no need for external energy.

3. A ball rolling uphill.

This is nonspontaneous, as it cannot happen on its own. A person would have to kick the ball for it to start moving uphill.

The second law of thermodynamics also deals with heat energy. If we were to place a metal rod (a material that conducts heat well) between a hot object (e.g, the ring on a stove) and a cold object (e.g, inside a glass of water) the heat will *spontaneously* move from the hot object to the cold.

This makes logical sense. Heat does not move from cold to hot, which is why your cup of coffee gets cold and why your soda gets warm over time.

It is *possible* to make heat move from the cold object to the hot object, but it would require energy to do so, making it nonspontaneous.

Natural spontaneous processes will lead to an increase in entropy, which can be expressed as: $\Delta S > 0$. Entropy will be greater than zero. We thus call spontaneous processes irreversible, as the changes caused by them in both the system and the environment cannot be changed back. For example, if you burn a five-dollar bill, you won't ever be able to buy a cheese burger from McDonald's with that bill, unless they accept ash as a payment method.

The only time entropy will not be greater than zero is in an ideal reversible process. What is a reversible process?

"A reversible process is a process in which the system and environment can be restored to exactly the same initial states that they were in before the process occurred, if we go backward along the path of the process" (Ling et al., 2016). Reversible processes require a quasi-static condition to work, meaning the process must happen slowly enough for the system to maintain equilibrium. A system can be restored to its original state quite easily, but having its environment restored at the same time is quite difficult. As an example, if an ideal gas expands into a vacuum and has double the volume that it had before, we can push it back into its original position with a piston, and then simply remove some heat from it to bring it back to its original temperature and pressure. The problem arises with the condition of the surroundings. While the gas may be the same as it was before expanding, we will change something in its surroundings, such as adding heat to it.

In an *ideal* reversible process, at best, $\Delta S = 0$. This is because in ideal reversible processes, entropy can be reversed.   But remember, this will require energy. All systems naturally (or spontaneously) become more disordered as time goes on.

Microstates are also a determining factor for the entropy of a system.

In English, *macro* refers to something large (or at least observable with the naked eye) and *micro* to something very small. It's important to realize, however, that macro and microstates in thermodynamics are not merely descriptions of a large and a small chemical system. Rather, they represent different approaches to understanding a system. (The macrostate always involves an amount of matter large enough for us to measure its volume, pressure, or temperature, i.e. in 'bulk'. But, in thermodynamics, a microstate is not just small amounts of matter; it is an analysis of how much energy molecules and other particles have.) A single microstate is one of a vast array of different accessible arrangements of the molecules' motional energy for a particular macrostate (Lambert & LibreTexts, 2018).

A macrostate does not change over time if its observable properties do not change.

Microstates, on the other hand, are all about the energy of the molecules in a system as well as the change across time. In a system, "energy is constantly being redistributed among its particles. In liquids and gases, the particles themselves are constantly redistributing in location as well as changing in the quanta (the individual amount of energy that each molecule has) due to their incessantly colliding, bouncing off each other with (usually) a different amount of energy for each molecule after the collision. Each specific way, each arrangement of the energy of each molecule in the whole system at one instant is called a microstate" (Lambert & LibreTexts, 2018).

We can then define a single microstate as something like "a theoretical absolutely instantaneous photo of the location and momentum of each molecule and atom in the whole macrostate" (Lambert & LibreTexts, 2018).

This is due to a "classical mechanics" approach, which assumes that molecules have a location and momentum. It is not possible to describe the behavior of molecules except through their energies at particular levels of energy in quantum mechanics. This is the more modern view of molecular behavior that we will use when discussing microstates.

The average molecule moves at around a thousand miles an hour, and a single one of these molecules collides with another about seven times in a billionth of a second. In the next one trillionth of a second, the system immediately changes to another microstate. Considering a mole of molecules (6×10^{23}) traveling at a very large number of different speeds, collisions (and thus changes in energy) of trillions of molecules occur in far less than a trillionth of a second (Lambert & LibreTexts, 2018).

Let's take a look at an overly simplified example of this.

Say you have two containers (or systems), one called A and the other B, both at a perfect vaccuum. They are linked by a sterile tube or pipe, and the whole system is at room temperature.

If you were to put molecules in container A, what would happen to them over time? For the sake of simplicity, let's say you placed four molecules in container A. According to the second law of thermodynamics, a spontaneous process will occur in which the molecules will move down their concentration gradient toward container B.

Let's label the molecules as 1, 2, 3, and 4. If you have molecules 1 and 2 in container A, molecules 3 and 4 will be in container B. If you have molecules 2 and 4 in container A, 1 and 3 will be in container B, and so on. This is entropy at work. Thus, the more possible microstates a system has, the higher its entropy (S) will be. The microstate that is most probable, or has the highest entropy, will be the one that the system is most likely to become over time (The Organic Chemistry Tutor, 2017).

The Third Law of Thermodynamics

Using the 3rd law of thermodynamics, we can essentially determine the amplitude of absolute entropy. At absolute zero (zero Kelvin), a 100% pure crystalline structure will have no entropy (S) (Mersha & Chemistry LibreTexts, 2013).

In simpler terms, its essential statement is about "the ability to create an absolute temperature scale, for which absolute zero is the point at which the internal energy of a solid is precisely 0" (Helmenstine, 2019b).

There are 3 potential formulations (or 3 "sub-laws," if you will) that accompany the third law. They are as follows:

1. "It is impossible to reduce any system to absolute zero in a finite series of operations.
2. The entropy of a perfect crystal of an element in its most stable form tends to zero as the temperature approaches absolute zero.
3. As temperature approaches absolute zero, the entropy of a system approaches a constant" (Helmenstine, 2019b).

All 3 formulations of the third law result in the same conclusion, depending upon how much you consider:

The third formulation contains the fewest constraints, stating only that entropy goes to a constant (the 'constant' referring to zero entropy as stated in formulation 2). A physical system, however, will never be able to decrease to 0 entropy due to quantum constraints. It is thus impossible to reduce a physical system on a finite level to absolute zero (which is what formulation 1 states).

For a structure that is not entirely crystalline, we could not say that there is no entropy in space, even though there would only be an extremely small amount of disorder (entropy). Also, while there is still some atomic motion present at zero Kelvin, we must assume that absolutely no entropy whatsoever exists at zero Kelvin in order to keep making sense of this world (Mersha & Chemistry LibreTexts, 2013).

Equilibrium

Equilibrium is the condition required for a reversible chemical reaction to occur, meaning that there is no net change in the amounts of reactants and products. As we have mentioned in dissimilar terms, a reversible chemical reaction is one in which the products, as soon as they are formed, react to produce the original reactants (Encyclopedia Britannica, 2016).

In chemical equations, equilibrium is expressed as the symbol "$\rightleftharpoons$" and is fairly simple to remember because of it. If "$\rightarrow$" denotes a chemical reaction, then logically "$\rightleftharpoons$" denotes a reaction that occurs but can also be reversed. In other words, the rate of the forward reaction is equal to the rate of the reverse reaction.

For example, if A turns into B during a reaction in equilibrium, B will turn back into A.

$A \rightleftharpoons B$

$Rate_F = Rate_R$

Once R_F (the forward rate, or the rate at which reactions occur) is equal to R_R (the reverse rate, or the rate at which products break down into their reactants), we know that we have a form of dynamic equilibrium. The concentration of both A and B becomes constant in equilibrium. It is important to note, however, that even though no change occurs during the reaction, there is still a process that occurs, and we thus do not consider equilibrium to be static.

To illustrate: Imagine you have a pair of ponds, each with a dragonfly population. Pond A has 10 dragonflies, and pond B has 30. If dragonflies move from pond A to pond B at a rate of 3 dragonflies a day, and the same amount per day from pond B to pond A, what is the change of dragonfly population in the ponds?

None. Because as dragonflies leave pond A, the exact same amount arrive from pond B, and vice versa. Even though there is movement and energy expended, there is no net change. The reaction, then, never stops. Let's take a look at an example.

N_2 (nitrogen gas) and H_2 (hydrogen gas) combine to create NH_3 (ammonia) during a *forward* process.

$$N_2(g) + 3H_2(g) \rightarrow 2NH_3(g)$$

However, as N_2 and H_2 combine, their concentration drops, and thus the number of collisions (and thus reactions) begins to decrease, and so does the rate at which NH_3 is produced. This is when the forward rate (R_F) slows down, as the frequency of forward reactions is decreasing. As this happens, the amount of NH_3 increases, meaning that as time goes on the amount of reverse reactions increases due to the fact there are more opportunities for the NH_3 molecules to break back down again into their reactants. This then logically means that the reverse rate

(R_R) speeds up as the concentration of NH3 increases. You can think of this as:

R_F increases with the concentration of reactants.

R_R increases with the concentration of products.

Eventually, the concentration of the reactants and the concentration of the products will meet a mutual pace of occurrence. When this happens, reactants (in our example, N2 and H2) are forming products (in our example, NH3) at the same rate at which the products are breaking back down into reactants. When this happens, we have reached equilibrium. You can also think of equilibrium as the balance between R_F and R_R.

Once we reach equilibrium, our equation changes:

$$N2(g) + 3H2(g) \rightarrow 2NH3(g)$$

to

$$N2(g) + 3H2(g) \rightleftharpoons 2NH3(g)$$

Now that we know what causes equilibrium to occur and how to identify it, what causes a system to become unbalanced, or to fall out of equilibrium?

The answer to this question was summarized by the French chemist Henry Louis Le Chatelier, and later named Le Chatelier's principle, which states that "if a dynamic equilibrium is disturbed by changing the conditions, the position of equilibrium shifts to counteract the change to reestablish an equilibrium. If a chemical reaction is at equilibrium and experiences a change in pressure, temperature, or concentration of products or reactants, the equilibrium shifts in the opposite direction to offset the change" (Chemistry Libretexts, 2020).

Change in concentration of one or more substance and changes in temperature or pressure can throw of chemical equilibrium. When this happens, either more reactions begin to occur, or more products begin to break down into their reactants. When more products begin to form, we say that the change shifts the reaction to the right. When more products begin to break down, we say the change shifts the reaction to the left (CrashCourse, 2013).

Let us go back to our example equation to take a look at how these changes affect the reactions that occur.

Increasing the concentration of our reactants, N_2 and H_2, will temporarily throw our system out of equilibrium, as there will be more products (NH_3) being formed than those breaking down. We can then say that the reaction has shifted to the right. However, the reactions within the system will return to equilibrium as the amount of NH_3 increases and begin to break down at the same rate at which they are being formed.

If you were to increase the concentration of products, the reverse would happen. Yes, the reactions within the system would be thrown out of equilibrium. But the reaction would shift to the *left*, meaning more NH_3 molecules would be breaking down than forming. Once again, the reactions in the system would return to equilibrium over time.

Increasing the pressure has a similar effect, though it is more permanent as long as the change in pressure is permanent. If we were to increase the pressure in a system, the reactants, N2 and H2, would be forced closer together and would thus create more NH3 molecules. The reaction would shift to the *right* (CrashCourse, 2013).

If we decrease the pressure, the inverse will occur. Molecules would become less compact, and thus fewer reactions would occur. However, because the NH3 molecules do not need to be close together to break down, the rate at which they do so would increase as the rate of reactions decreases. The reaction would shift to the *left*. Take note that the effects of pressure change primarily apply to *gases* (CrashCourse, 2013).

How about temperature? Remember that endothermic reactions (which absorb heat) favour heat and exothermic reactions (which release heat) prefer cold. Adding heat to an exothermic reaction will cause the reaction to shift to the left, and cooling it will cause the reaction to proceed to the right.

Temperature changes are caused by increasing or decreasing heat flow. This will then "shift chemical equilibria toward the products or reactants, which can be determined by studying the reaction and deciding whether it is endothermic or exothermic" (Pulido et al., 2013).

In endothermic reactions, ($\Delta H > 0$) thermal energy is absorbed via the reaction. Endothermic reactions can also be viewed as those that require more thermal energy to overcome the forces of attraction between molecules (the activation energy) than that released when new bonds are formed (Pulido et al., 2013).

In exothermic reactions, (ΔH<0) thermal energy is general with reactions. When new bonds are generated during reactions, thermal energy—which is required to break bonds in the reactants—is released.

We classify the forward reaction as being exothermic, as the combination of reactants to form products releases energy. On the right side of the equation, heat is observed as a result of the net difference between the energy required to break the bonds on the left side of the equation and the energy released to form the product(s) (Pulido et al., 2013).

What is fascinating is how these balance each other out once again. As endothermic reactions occur at higher temperatures, heat is absorbed and will thus cool down the rest of the reactants and products and slow down other endothermic reactions, meaning exothermic reactions will then begin to take place and slowly but surely bring back balance in the form of equilibrium by releasing heat (CrashCourse, 2013).

Practice Questions

1. **Question: "Identify the oxidation state of the atoms in the following compounds"** **(Chemistry LibreTexts, 2021):**

   A) PCl_3

   B) CO_3^{2-}

C) H2S

Answers:

"A) PCl3

Chlorine has an oxidation state of -1. There are 3 chlorine atoms present in this compound, and so chlorine has a total oxidation state of (-1)3 = -3. Because this is a neutral compound, we know that the net charge of the molecule must equal 0 (Rule #4). Therefore, we can find the oxidation state of phosphorus because :

O.S.(P)+O.S(Cl)=net charge

O.S.(P)+(-3)=0

O.S.(P)=+3

The oxidation state of chlorine atom =-1, total oxidation state of Chlorine atoms: -3

The oxidation state of phosphorous atom=+3

B) CO32−

According to Rule #2, oxygen has an oxidation number of -2. Since there are 3 oxygen atoms, the overall oxidation state of the oxygens is (-2)3=-6. Because the overall molecule has a net charge of 2- , carbon must have an oxidation state of (-2)-(-6)=+4.

Oxidation state of oxygen atom=-2

Oxidation state of carbon atom=+4

C) H2S

According to Rule #3, hydrogen has an oxidation state of +1. Since there are two hydrogen atoms, the total oxidation state of the hydrogens is (+1)2=+2. The overall molecule has a net charge of 0, and so the oxidation state of the sulfur atom is 0-(+2)=-2.

Oxidation state of hydrogen atom=+1

Oxidation state of sulfur atom=-2

Explanation and Rules

An atom's oxidation state is representative of the number of electrons that a specific atom has lost or gained when binding to another atom. This allows us to determine the changes that are caused in redox reactions as well as balance redox reactions. Below are some of the rules for determining an atom's oxidation number:

The oxidation state of an element that is not combined with another element, Fe, H2, O2, P4, S8 is zero (0).

The oxidation state of oxygen in a compound is -2, except for when it is in peroxides like H2O2, and Na2O2—in this case the oxidation state for O is -1.

The oxidation state of hydrogen is +1 in its compounds, except for metal hydrides such as NaH and LiH, where the oxidation state for H is -1.

The net charge on a molecule or ion is equal to the algebraic sum of the oxidation states of all the atoms present in the species.

> +1 for alkali metals (Group 1): (Li, Na, K, Rb, Cs)

> +2 for alkaline earth metals (Group 2): (Be, Mg, Ca, Sr, Ba)

> The oxidation number of fluorine is always −1. Chlorine, bromine, and iodine usually have an oxidation number of −1, unless they are combined with an oxygen or fluorine atom.

The oxidation number of a monatomic ion is equal to the charge on the ion" (Chemistry LibreTexts, 2021).

2. Question & Additional Learning: Why might an electrochemical reaction that is thermodynamically favored (or spontaneous) require an overvoltage to occur?

Answer: "Overvoltage occurs when there is a requirement for additional voltage to be added to an electrolytic cell in order for the rate of reaction to occur as it would in an ideal setting. Although a reaction may already be thermodynamically favored prior to the additional voltage, there are a few reasons why it may still be necessary. First, although a reaction may be thermodynamically favored and thus exothermic, it may be advancing at a less than optimal rate due to high activation energy. Overvoltage can aid in lowering the activation energy and thus increasing the rate, all while the reaction remains thermodynamically favored. Additionally, in some reactions in which gases are produced, the flow of electrons is slowed by the gases being formed. Once again, overvoltage can alleviate the slower rate even though the reaction was already favorable" (Chemistry LibreTexts, 2021).

3. Question: Why is it preferable to use mixtures of molten salts instead of pure salt during electrolysis?

Answer: "Mixtures of molten salts are used over pure salts during electrolysis because they have good electric conductivity, a higher heat capacity, better thermal conductivity, and can act as a solvent. The pure salts have higher melting points alone, but when they are mixed, their combined melting point (known as the eutectic point) is lower. This helps the reaction move forward, making electrolysis easier" (Chemistry LibreTexts, 2021).

4. Question: Is it correct to say that the reaction has 'stopped' when it has reached equilibrium? Explain your answer and support it with a specific example.

Answer: "It is not correct to say that the reaction has 'stopped' when it has reached equilibrium because it is not necessarily a static process where it can be assumed that the reaction rates cancel each other out to equal zero or be 'stopped', but rather a dynamic process in which reactants are converted to products at the same rate products are converted to reactants. For example, a soda has carbon dioxide dissolved in the liquid, and carbon dioxide gas under pressure between the liquid and the cap that is constantly being exchanged with each other. The system is in equilibrium and the reaction taking place is:

$$CO_2(g) + 2H_2O(l) \rightleftharpoons H_2CO_3(aq)"$$ (Chemistry LibreTexts, 2021).

5. Question: A gas in a closed container is heated, causing the lid of the container to rise. The gas performs 3J of work to raise the lid, such that it has a final total energy of 15J. How much heat energy was added to the system?

Possible Answers:

18J

5J

12J

50J

45J

Answer: 18J

Explanation:

For this problem, use the first law of thermodynamics. The change in energy equals the increase in heat energy plus the work done.

$$\Delta U = Q + W$$

We are given the amount of work done by the gas and the total energy of the system. Using these values, we can solve for the heat added.

$$\Delta U = Q + W$$

$$15J = Q - 3J$$

$$Q = 18J$$

6. Question: Which is not characteristic of an endothermic reaction?

"Possible Answers:

Releasing energy to surroundings.

Breaking of a chemical bond.

Absorption of energy from surroundings.

The net change in enthalpy is positive.

A chemical reaction that feels cold" (GRE Chemistry Varsity Tutors, n.d.).

Answer: Releasing energy to surroundings.

Explanation:

"An endothermic reaction absorbs energy in the form of heat. An endothermic reaction involves the breaking of chemical bonds because it involves the absorption of energy. Endothermic reactions tend to feel cold because it is taking heat away from your skin. Since endothermic reactions involve absorbing energy, often in the form of heat, the change in enthalpy is positive.

Therefore, the answer is "releasing energy to surroundings" which is not a characteristic of an endothermic process. Instead, it is a characteristic of an exothermic reaction, which is the direct opposite of an endothermic reaction" (GRE Chemistry Varsity Tutors, n.d.).

7. Question: "Which of the following reactions has a positive value for change in entropy ($\Delta S > 0$)?

Possible Answers:

$N_2(g) + 3H_2(g) \rightarrow 2NH_3(g)$

Water freezing into ice

Water evaporates into water vapor

A salt precipitates out of a saturated solution

Answer: Water evaporates into water vapor" (GRE Chemistry Varsity Tutors, n.d.).

Explanation:

"Entropy is a measure of the disorder of a system. In other words, when the components of a system become more uniform and spread out, the entropy of the system has increased. For example, when a liquid becomes gaseous, the molecules separate from one another, increasing the disorder of the system. As a result, the evaporation of water results in an increase in entropy.

All other options result in a more ordered system, which decreases entropy" (GRE Chemistry Varsity Tutors, n.d.).

8. Question: "Which law of thermodynamics states that a crystal's entropy at a temperature of absolute zero is equal to 0?" (GRE Chemistry Varsity Tutors, n.d.).

Possible Answers:

First law

Zeroth law

Third law

Second law

Answer: Third law

Explanation:

"There are four laws of thermodynamics:

Zeroth law of thermodynamics: If two independent systems are each in thermodynamic equilibrium with a thrid independent system, then the first two systems must be in thermodyanmic equilibrium with each other.

First law of thermodynamics: Energy is neither created nor destroyed.

Second law of thermodynamics: The entropy of the universe increases following every reaction.

Third law of termodynamics: At a temperature of 0 Kelvin (absolute zero), the entropy of the crystal is 0" (GRE Chemistry Varsity Tutors, n.d.).

9. Question: What is the importance of the zeroth law of thermodynamics?

Answer: "Zeroth law of thermodynamics states that if two bodies are in thermal equilibrium with a third body, then they are in thermal equilibrium with each other. Thermal equilibrium means the same temperature. When two bodies are in thermal equilibrium, then there will be no heat transfer between them. So, the zeroth law of thermodynamics is used for temperature measurements" (PSIBERG Team, 2022).

10. Question: A glass of cold water is placed in a sealed room. After an infinite amount of time, what will happen?

"Possible Answers:

The water will warm to the temperature of the air.

Both the water and the air will remain the same temperature, unless an outside force is applied.

The water will warm and the air in the room will cool, until they are the same temperature.

The air will cool to the temperature of the water.

There is insufficient information to solve.

Answer: The water will warm and the air in the room will cool, until they are the same temperature.

Explanation:

The second law of thermodynamics states that closed systems constantly move towards a state of thermal equilibrium. Since we are looking at a closed system, that means we must be moving towards a state of equilibrium. The only way that happens is if the air cools and the water warms, until they both reach a new final temperature.

Looking at this question in terms of heat transfer, we can infer that the glass of water will warm up, but there must be a transfer of heat to the glass in order for this to occur. The heat comes from the air, causing it to cool as the glass warms" (High School Varsity Tutors, n.d.).

11. Question: Of these elements, which would you expect to be easiest to reduce: Se, Sr, or Ni? Explain your reasoning.

Answer: Remember the pneumonic, OIL RIG: "Oxidation is loss, Reduction is gain." When an element is reduced, it gains electrons.

Elements want to have a full valence shell (the outermost shell that accounts for the reactivity of the element). Since Se is a Group 6 element, it has 6 valence electrons. The most stable state has 8 valence electrons (a full shell). Se would be the easiest to reduce because it readily accepts electrons to complete its valence shell.

Sr is in Group 2, meaning it has 2 valence electrons. Elements in Group 1 or Group 2 are more likely to give up electrons to empty their outermost shell. In this case, those elements would be oxidized because they are giving up electrons.

Ni is part of the transition metals. Most transition metals have a lower number of valence electrons. In this case, nickel has 2 valence electrons, making it more likely to be oxidized like Sr" (Chemistry LibreTexts, 2021).

12. Question: Which of the following statements is correct?

"The presence of reacting species in a covered beaker is an example of an open system.

There is an exchange of energy as well as matter between the system and the surroundings in a closed system.

The presence of reactants in a closed vessel made of copper is an example of a closed system.

The presence of reactants in a thermos flask or any other closed insulated vessel is an example of a closed system.

Answer: The presence of reactants in a closed vessel made of copper is an example of a closed system.

Explanation: There is no exchange of matter in a closed system (for example, the presence of reactants in a closed vessel made of conducting material, such as copper), but there is an exchange of energy between the system and its surroundings" (Byju's, n.d.).

13. Question: How do we know if a reaction is spontaneous?

Possible Answers:

The value of k (the rate constant) tells us.

The sign of $\Delta G°$.

The second law of thermodynamics.

The first law of thermodynamics.

The sign of ΔH.

Answer: The sign of $\Delta G°$

"Explanation: A reaction is spontaneous if and only if $\Delta G°<0$. The sign of ΔH only tells us if a reaction is exothermic or endothermic. The rate constant only tells us the rate at which a given chemical reaction proceeds based on the reactants and the products. The first law of thermodynamics only tells us that energy is conserved and can neither be created nor destroyed. The second law of thermodynamics states that the entropy of the universe is spontaneously increasing; however, under the right conditions, these reactions may proceed" (College Varsity Tutors, n.d.).

14. Question & Additional Learning:

"Suppose that a given chemical reaction has an enthalpy change of +175kJmol and an entropy change of +500Jmol·K. At what temperature would this chemical reaction be in equilibrium?" (College Varsity Tutors, n.d.).

Possible Answers:

200K

300K

250K

350K

400K

"Answer: 350K

Explanation:

For this question, we're asked to determine the temperature needed for a reaction to be in a state of equilibrium. We are also provided with the entropy and enthalpy changes for this reaction.

Remember that for a reaction to be in a state of equilibrium, the value of its change in free energy must be equal to zero. Furthermore, we can state the relationship between the change in free energy with the change in entropy, the change in enthalpy, and the temperature as follows.

$$\Delta G = \Delta H - T\Delta S$$

Because the ΔG term in the above expression must be equal to zero, we can plug this value in and then isolate the term for temperature.

$$0 = \Delta H - T\Delta S$$

$$T\Delta S = \Delta H$$

$$T = \Delta H / \Delta S$$

Now that we have this expression, we just need to plug in the two values given to us in the question stem. Remember to make the units match, though!

$$T = {}^{175kJ/mol}/_{0.5kJmol \cdot K} = 350K"\ \text{(College Varsity Tutors,}$$
n.d.).

15. Question: If we were to put Sodium Chloride (NaCl) through electrolysis in *water* using a battery and two electrodes (an anode and a cathode), what product would form at the *negative* electrode?

Possible Answers:

Oxygen

Chlorine

Hydrogen

Sodium

Answer: Hydrogen

Explanation:

The ions present in sodium chloride solution are:

Na+

Cl-

H+ (from water)

OH- (from water)

Positively charged ions (cations; Na+ and H+) will be attracted to the negative electrode. Due to hydrogen having a lower reactivity than sodium, it will be discharged at the electrode. Hydrogen ions will then gain electrons and form hydrogen gas, floating up and away from the container where the electrolysis is being conducted.

Chapter 6: Acid Base Chemistry

Acids and bases are sort of like two sides of the same coin in chemistry. They're linked, but they have differing traits that make them distinctly (and importantly) different.

Acids can be identified by searching for hydrogen atoms that, upon being placed into a solution, are released as hydrogen cations. All acids must contain at least 1 hydrogen atom. HCl and HF are examples of acids. There are other features that acids have that are higher up the analysis hierarchy.

"An acid in a water solution tastes sour (which is why lemon juice, a liquid mixture that is high in acid and is made up mainly of water, tastes sour), changes the colour of blue litmus paper to red, reacts with some metals (e.g., iron) to liberate hydrogen, reacts with bases to form salts, and promotes certain chemical reactions (acid catalysis)" (Percy Bell & Encyclopaedia Britannica, 2018).

Bases, on the other hand, typically contain a hydroxide ion. NaOH and KOH are examples of bases. Bases taste bitter and change the colour of red litmus paper to blue, the inverse of acids. When bases react with acids, they form salts and promote certain chemical reactions (base catalysis). When a molecule has a metal and a hydrogen atom, it is a base. When a molecule has a hydrogen atom that is bonded to a nonmetal, typically it would be an acid. Another way of distinguishing acids from bases would be the charge that the H atom has. If the hydrogen has a positive charge (H^+), it's an acid. If it has a negative charge

(H-), it's a base. Acids, in general, tend to be positively charged while the typical base has a negative charge.

An acid–base reaction is a type of chemical reaction that involves the exchange of one or more hydrogen ions, H^+, between neutral molecules, such as water, or electrically charged ions, such as ammonium, NH_4^+; hydroxide, OH^-; or carbonate, CO_3^{2-}. We also use the term in reference to similar processes that occur in molecules and ions that do not donate hydrogen ions but are still acidic (Percy Bell & Encyclopaedia Britannica, 2018).

Arrhenius Theory

To understand the classifications of acids and bases, we will be taking a look at the Arrhenius and the Brønsted-Lowry theories. For now, let us take a look at the first theory—the Arrhenius theory. In 1884, the Swedish chemist Svante Arrhenius proposed two specific classifications of compounds: acids and bases. His definition states that "an acid produces H+ in solution and a base produces OH-" (Broderick et al., 2019). This theory was developed by Arrhenius in 1887 (The Editors of Encyclopedia Britannica, 2016).

Regardless of whether a base or an acid is being dissolved, certain ions are released into the aqueous solution used to dissolve them. An acid that increases the concentration of H^+ ions present when added to water is called an Arrhenius acid. These H^+ ions will then form the hydronium ion (H_3O^+) when they combine with surrounding water molecules. An Arrhenius base will increase the amount of hydroxide ions when dissolved in a solution. We can represent this process using the following equation:

$$HCl(aq) \rightarrow H^+(aq) + Cl^-(aq)$$

Here we see the acid (HCl) being dissolved in water and broken down into its reactants, H^+ and Cl^-. To represent the reaction between an H^+ atom and a water molecule, we simply add H_2O to the reactants side:

$$HCl(aq) + H_2O(l) \rightarrow H_3O^+(aq) + Cl^-(aq)$$

There are limits to this theory, however. For example, the Arrhenius theory only works with acids and bases in aqueous solutions, and while it successfully describes how acids and bases react with each other to make water and salts, it doesn't provide an explanation for why some substances that do not contain hydroxide ions (e.g, F^- and NO_2^-) are able to make basic solutions in water. Fortunately, the Brønsted-Lowry definition of acids and bases answers that question.

Brønsted-Lowry Theory

Similar to Svante Arrhenius, Johannes Nicolaus Brønsted and Thomas Martin Lowry independently developed definitions of acids and bases around the year 1923. Their definitions, unlike Arrhenius', were based on the compounds' abilties to either donate or accept protons (H+ ions).

In this theory, acids are defined as proton donors and bases are defined as proton acceptors. You can already see how this broader definition is more all-encompassing than that of the Arrhenius theory—substances no longer need to be composed of H^+ OH^- ions in order to be classified as an acid or base. There are also some compounds that act as both a Brønsted-Lowry acid and base. Such compounds are known as being amphoteric (Broderick et al., 2019).

"Consider the following chemical equation:

$$HCl(aq) + NH3(aq) \rightarrow NH4 + (aq) + Cl^-(aq)$$

Here, hydrochloric acid (HCl) "donates" a proton (H^+) to ammonia (NH3) which 'accepts' it, forming a positively charged ammonium ion ($NH4^+$) and a negatively charged chloride ion (Cl^-). Therefore, HCl is a Brønsted-Lowry acid (donates a proton) while the ammonia is a Brønsted-Lowry base (accepts a proton). Also, Cl^- is called the conjugate base of the acid HCl and $NH4^+$ is called the conjugate acid of the base NH3" (Broderick et al., 2019).

Like in the Arrhenius theory, acids release protons, but unlike the Arrhenius theory, a base can *accept* a proton. A basic salt, such as Na^+F^-, generates OH^- ions in water by taking protons from the water itself (to make HF):

$$F(aq)^- + H2O(l) \rightleftharpoons HF(aq) + OH^-$$

Another similarity between an Arrhenius and a Brønsted acid is that when it dissolves, it increases the concentration of hydrogen (H^+) ions in the solution. But once again, the difference between these theories shines in the case of the base. A Brønsted base dissolves by taking a proton from the solvent (water) to generate a hydroxide ion (OH^-).

Acid Dissociation: $HA(aq) \rightleftharpoons A^-(aq) + H^+(aq)$

Acid Ionization Constant: $K_a = \dfrac{[A^-][H^+]}{[HA]}$

Base dissociation: $B(aq) + H2O(l) \rightleftharpoons HB^+(aq) + OH^-(aq)$

Base Ionization Constant: $K_b = \dfrac{[HB^+][OH^-]}{[B]}$

pH Levels

pH is a measure of how acidic/basic water is. Its value ranges from 0 all the way up to 14, with 7 being the middle ground, and logically, the neutral pH level. Acidity is indicated by pH levels of less than 7, whereas bases are indicated by pH levels greater than 7. Basically, pH measures the amount of hydrogen and hydroxyl ions present in the water being tested. Water that has a higher count of hydrogen ions is considered acidic, and water that has more free-moving hydroxyl ions is basic (a base). The pH of water can be affected by chemicals in it, so it is an important indicator of chemical changes in the water. pH is reported in "logarithmic units," and each number represents a 10-fold change in how acidic/basic the water being tested is. For example, if you had water with a pH of five, it would be ten times more acidic than water that had a pH of six (U.S Department of the Interior & Water Science School, 2019).

pH has many uses, including whether or not flora and fauna can survive in water (if you've ever kept tropical fish, you would know this), whether it is safe for human consumption, and even the water's solubility. "The pH of water determines the solubility (amount that can be dissolved in the water) and biological availability (amount that can be utilized by aquatic life) of chemical constituents such as nutrients (phosphorus, nitrogen, and carbon) and heavy metals (lead, copper, cadmium, etc.). For example, in addition to affecting how much and what form of phosphorus is most abundant in the water, pH also determines whether aquatic life can use it. In the case of heavy metals, the degree to which they are soluble determines their toxicity. Metals tend to be more toxic at lower pH because they are more soluble" (U.S Department of the Interior & Water Science School, 2019).

Acid-Base Indicators

Have you ever dyed a piece of clothing? If you have, you will have a good idea of what a macroscale representation of an indicator looks like.

An acid-base indicator is the most common way to determine the pH of a solution. Indicators are large organic molecules that work somewhat like color dyes. In contrast to most dyes that do not change color with the amount of acid or base present, there are many molecules, known as acid-base indicators, that do change color with changes in hydrogen ions. These molecules are generally weak acids themselves.

The most commonly used of these indicators is litmus paper. Litmus paper turns red when exposed to water with a pH level equal to or below 4.5, and turns blue when exposed to water with a pH level equal to or above 8.2. Litmus paper itself is either red or blue, and when exposed to an acid (pH<4.5), blue paper turns red, and when exposed to a base (pH>8.2), red paper turns blue. Red paper remains the same when exposed to an acid, and the same goes for blue paper when exposed to a base.

The development of biodiesel fuels is a major focus of current research. Often, vegetable oil is treated with lye to create the biofuel. Lye (Sodium Hyrdroxide) is a white crystalline solid that contains the Na^+ (sodium) cation and the OH^- (hydroxide) anion. It readily absorbs moisture until it dissolves. It is a highly corrosive base that, in the production of fuel, removes organic and sulfuric acids found within the oils being treated (Encyclopaedia Britannica & Gregersen, 2022).

The amount of organic/sulfuric oils that need to be removed is a variable that needs to be determined so that those refining the oil will know how much lye to add to make the final fuel. It is necessary to titrate the native vegetable oil before adding lye so the level of free acid is determined. Afterward, the amount of lye added can be adjusted based on the amount needed for the neutralization of the acids (Chemistry LibreTexts & CK-12, 2016b).

"In the neutralization of hydrochloric acid by sodium hydroxide, the mole ratio of acid to base is 1:1.

$HCl(aq) + NaOH(aq) \rightarrow NaCl(aq) + H_2O(l)$

One mole of HCl would be fully neutralized by one mole of NaOH. If, instead, hydrochloric acid was reacted with barium hydroxide, the mole ratio would be 2:1.

$2HCl(aq) + Ba(OH)_2(aq) \rightarrow BaCl_2(aq) + 2H_2O(l)$

Now two moles of HCl would be required to neutralize one mole of Ba(OH)$_2$" (Chemistry LibreTexts & CK-12, 2016b).

The mole ratio means there will be an equal number of H^+ ions supplied by the acid and OH^- ions supplied by the base, which is the condition required for neutralization to occur. In a neutralization reaction, the equivalence point is the point where the number of hydrogen ions is equivalent to the number of hydroxide ions.

Oftentimes, testing the concentration of an acid or a base is imperative. Quite literally, it can be the difference between life and death. So, having a way to determine the concentration of an acid or base through experimentation and testing is not only useful, but important. We do this through the use of controlled neutralization reactions.

A titration is an experiment where a volume of a solution of known concentration is added to a volume of another solution in order to determine its concentration by observing the changes and reactions that take place. Though there are several types of titrations, the main titration type used is acid-base neutralization reactions.

As a chemist, it is important to have a visual representation of whether or not the neutralization reaction has taken place when performing an acid-base titration. This takes us back to the use of indicators. As litmus is commonly used for weaker acids and bases, a commonly used indicator for strong acid-strong base titrations is phenolphthalein.

We use the ionization of acids and bases in water to determine whether or not they are strong or weak. The stronger an acid is, the greater its ionization in an aqueous solution. An example of this is hydrogen chloride (HCl), which ionizes completely into hydrogen ions and chloride ions when placed in water (Chemistry LibreTexts & CK-12, 2016a). This can be expressed in the following equation:

$HCl(g) \rightarrow H^+(aq) + Cl^-(aq)$

A weak acid, on the other hand, will only ionize slightly in an aqueous solution. Acetic acid (an acid that makes up vinegar) is an example of a weak acid. Its ionization can be expressed in the following equation:

$CH_3COOH(aq) \rightleftharpoons H^+(aq) + CH_3COO^-(aq)$ (Chemistry LibreTexts & CK-12, 2016a).

When phenolphthalein is added to a solution, it turns from colorless to brilliant pink as it changes from acidic to basic.

The basic steps in a titration reaction are as follows:

> "A measured volume of an acid of unknown concentration is added to an Erlenmeyer flask.

Several drops of an indicator are added to the acid and mixed by swirling the flask.

A burette is filled with a base solution of known molarity.

The stopcock of the burette is opened and a base is slowly added to the acid, while the flask is constantly swirled to ensure mixing. The stopcock is closed at the exact point at which the indicator changes color.

The standard solution is the solution in a titration whose concentration is known. In the titration described above, the base solution is the standard solution. It is very important in a titration to add the solution from the burette slowly so that the point at which the indicator changes color can be determined accurately. The end point of a titration is the point at which the indicator changes color. When phenolphthalein is the indicator, the end point will be signified by a faint pink color" (Chemistry LibreTexts & CK-12, 2016b).

Practice Questions

1. **Question: Identify the conjugate acid-base pairs in the following equilibrium.**

   $HSO_4^-(aq) + H_2O(l) \rightleftharpoons SO_4^{2-}(aq) + H_3O^+(aq)$

   Answer: $HSO_4^-(aq)_{[acid]}$ + $H_2O(l)_{[base]}$ $\rightleftharpoons$ $SO_4^{2-}(aq)_{[conjugate\ base]}$ + $H_3O^+(aq)_{[conjugate\ acid]}$

2. **Question: Show NH3 as both a conjugate acid and a conjugate base in equation form.**

   Answer: An example for NH3 as a conjugate acid: $NH_2^- + H^+ \rightarrow NH_3$; as a conjugate base: $NH_4^+(aq) + OH^-(aq) \rightarrow NH_3(aq) + H_2O(l)$

3. **Question: Show by suitable net ionic equations that the following species can act as a Brønsted-Lowry acid: H3O⁺**

   Answer: $H_3O^+(aq) \rightarrow H^+(aq) + H_2O(l)$

4. **Question: Question: Show by suitable net ionic equations that the following species can act as a Brønsted-Lowry acid: HCl**

   Answer: $HCl(l) \rightarrow H^+(aq) + Cl^-(aq)$

5. **Question: Question: Question: Show by suitable net ionic equations that the following species can act as a Brønsted-Lowry acid: NH3**

   Answer: $NH_3(aq) \rightarrow H^+(aq) + NH_2^-(aq)$

6. **Question: What are amphiprotic species? Illustrate with suitable equations.**

   Answer: Amphiprotic species may either gain or lose a proton in a chemical reaction, thus acting as a base or an acid. An example is H2O.

   As an acid: $H_2O(aq) + NH_3(aq) \rightleftharpoons NH_4^+(aq) + OH^-(aq)$.

As a base:

$$H_2O(aq) + HCl(aq) \rightleftharpoons H_3O^+(aq) + Cl^-(aq)$$

7. **Question: Predict which compound in the following pair of compounds is more acidic and explain your reasoning.**

HSO$_4^-$ or HSeO$^-_4$

Answer: HSO$_4^-$; higher electronegativity of the central ion.

8. **Question: Predict which compound in the following pair of compounds is more acidic and explain your reasoning.**

NH$_3$ or H$_2$O

Answer: NH$_3$ is a base and water is neutral, or decide on the basis of Ka values.

9. **Question: Explain why equilibrium calculations are not necessary to determine ionic concentrations in solutions of certain strong electrolytes such as NaOH and HCl.**

Answer: Strong electrolytes are 100% ionized, and, as long as the component ions are neither weak acids nor weak bases, the ionic species present result from the dissociation of the strong electrolyte.

10. **Question: Under what conditions are equilibrium calculations necessary as part of the determination of the concentrations of all**

ions of some other strong electrolytes in solution?

Answer: Equilibrium calculations are necessary when one (or more) of the ions is a weak acid or a weak base.

11. **Question: Explain how to choose the appropriate acid-base indicator for the titration of a weak base with a strong acid.**

Answer: At the equivalence point in the titration of a weak base with a strong acid, the resulting solution is slightly acidic due to the presence of the conjugate acid. Thus, pick an indicator that changes color in the acidic range and brackets the pH at the equivalence point.

12. **Question: The density of liquid water decreases as the temperature increases from 25°C to 50°C. Will this effect cause Kw (ion-product constant of water) to increase or decrease?**

Answer: This will affect Kw as it is dependent on temperature. As the temperature increases, an endothermic process occurs (energy must be absorbed to break the bonds). Consequently, according to Le Chatelier, an increase in temperature favors the forward reaction thus the position of equilibrium shifts toward the right-hand side and Kw becomes larger.

13. **Question: Identify the following as either a Lewis Acid or a Lewis Base: NH3**

Answer: NH3 is a lewis base because nitrogen has a lone pair of electrons to "donate."

14. Question: Identify the following as either a Lewis Acid or a Lewis Base: Ag⁺

Answer: Ag⁺ is a lewis acid because it has an unfilled octet and thus is able to accept a pair of electrons.

15. Question: Identify the following as either a Lewis Acid or a Lewis Base: H2O

Answer: H2O is a lewis base because oxygen has two lone pairs of electrons to "donate."

Chapter 7: Organic Chemistry

When we talk about organic chemistry, it is important to distinguish it from the other chemistry types. One can mistakenly refer to all types of chemistry as being organic, but when we refer to organic chemistry itself, we are referring to "the study of the structure, properties, composition, reactions, and preparation of carbon-containing compounds. Most organic compounds contain carbon and hydrogen, but they may also include any number of other elements (e.g., nitrogen, oxygen, halogens, phosphorus, silicon, sulfur)" (American Chemical Society, n.d.).

The study of organic chemistry was initially limited to the study of compounds produced by living organisms (hence the name and the possible confusion with the study of many types of chemistry, as organic beings are capable of a variety of chemical changes and reactions). Today, we have broadened the subject of organic chemistry to include human-made substances such as plastics (American Chemical Society, n.d.). Organic chemistry could be considered less sciencey and more artful than the other types, as it is considered to be the most creative of the chemical sciences. It allows chemists to explore molecules and compounds and create new ones, as opposed to sticking to hard and fast rules as the other chemistry types often do. Organic chemists will often spend a lot of time developing new compounds and finding better ways of synthesizing existing ones (American Chemical Society, n.d.). Organic compounds exist pretty much everywhere. For example, modern materials—the stuff your chair, house, local store, and road are made of—are likely going to be partially composed of organic compounds. Biochemistry, biotechnology, and medicine are all based on them, so they are vital to economic growth. There are a number of places where organic compounds can be found. These include agrichemicals, coatings, cosmetics, detergents, dyestuff, food, fuels, petrochemicals, pharmaceuticals, plastics, and rubber (American Chemical Society, n.d.).

The first of the organic chemistry topics that we will be considering is the most basic and most common of the organic compounds: the hydrocarbons.

Hydrocarbons

Hydrocarbons are the most widespread of the organic chemical compounds. They are made up of all organic chemical compounds composed of the elements carbon (C) and hydrogen (H). The carbon atoms create the framework of these compounds, and the hydrogen atoms attach to that framework to form different configurations.

These different hydrocarbon configurations are the fundamental constituents of petroleum and natural gas. They make up the compounds and mixtures found in the raw materials for the production of plastics, fibers, rubber, solvents, explosives, and industrial chemicals. Not only are they common in man-made structures, but many of them are found throughout nature as well.

"In addition to making up fossil fuels, they (hydrocarbons) are present in trees and plants." An example of these can be found in "the form of pigments called carotenes that occur in carrots and green leaves. More than 98 percent of natural crude rubber is a hydrocarbon polymer, a chainlike molecule consisting of many units linked together. The structures and chemistry of individual hydrocarbons depend in large part on the types of chemical bonds that link together the atoms of their constituent molecules" (Carey & The Encyclopaedia Britannica, 2018).

There are two types of hydrocarbons: aliphatic hydrocarbons and aromatic hydrocarbons. An aliphatic hydrocarbon is one containing chains of carbon atoms (Agnew et al., 2019). The name aliphatic comes from the Greek word *aleiphar*, meaning 'fat', and was named that way to describe the hydrocarbons derived by chemical degradation of fats or oils. As for the related aromatic hydrocarbons, certain pleasant-smelling plant extracts are chemically degraded to produce them. Modern terminology makes use of the terms aliphatic and aromatic, but distinguishes between the compounds they describe more on the basis of structure than origin (Carey & The Encyclopaedia Britannica, 2018).

There are 3 types of aliphatic hydrocarbons: Alkanes, alkenes, and alkynes. They are distinguished by the number of bonds found within them; alkanes contain only single bonds, alkenes have a carbon-carbon double bond, and alkynes have a carbon-carbon triple bond (Carey & The Encyclopaedia Britannica, 2018). At times, we come across cycloalkanes (or cycloalkenes or cycloalkynes), an aliphatic hydrocarbon with a ring of C atoms. Alkanes are also called saturated hydrocarbons as they have the maximum number of H atoms possible according to the rules of covalent bonds (Agnew et al., 2019).

On the other hand, aromatic carbons—also known as arenes—are those that are significantly more stable than they may seem when looking at their Lewis-dot structures. They possess a sort of "special stability" (Carey & The Encyclopaedia Britannica, 2018). They contain a benzene ring as a structural unit, or non-benzenoid aromatic hydrocarbons, which possess this special stability but lack a benzene ring as a structural unit. Benzene, as well as many other aromatic compounds, contains hexagonal unsaturated rings of six carbon atoms. A non-benzenoid compound that exhibits aromatic behavior is called a non-benzenoid aromatic compound, and is void of a benzene nuclei. Non-benzenoid aromatic compounds can often have one or more rings fused, but none of them will be a benzene ring, even if the number of rings present goes above 5 (Jha, 2020).

Arenes are cyclic unsaturated hydrocarbons whose chemical properties differ so sharply from those of conjugated alkenes (polyenes) that they are usually considered a separate class of hydrocarbons (Robert et al., 2014)

"This classification of hydrocarbons serves as an aid in associating structural features with properties but does not require that a particular substance be assigned to a single class. Indeed, it is common for a molecule to incorporate structural units characteristic of two or more hydrocarbon families. A molecule that contains both a carbon-carbon triple bond and a benzene ring, for example, would exhibit some properties that are characteristic of alkynes and others that are characteristic of arenes" (Carey & The Encyclopaedia Britannica, 2018).

Alkanes

Let us take a moment to consider each of the subcategories of hydrocarbons.

Alkanes can most simplistically be described as hydrocarbon chains. All of them are made up of varying amounts of carbon and hydrogen atoms that can either form a single unbranched chain, or the primary chain of carbon atoms can have one or more shorter chains that form branches. Alkanes only contain *single* bonds between hydrogen and carbon atoms or between two carbon atoms. Alkanes that contain four or more carbon atoms are multi-organizational and can have more than one arrangement of atoms (Chemistry LibreTexts, 2014).

For example, butane (or n-butane, C_4H_{10}) is normally written as $CH_3CH_2CH_2CH_3$, and the carbon atoms form a single unbranched chain. However, it can also be written as $CH_3(CH_2)_2CH_3$. We call this the condensed structural formula for isobutane. The primary chain of the 3 carbon atoms in isobutane has a 1-carbon chain branching at the central carbon (Chemistry LibreTexts, 2014).

The easiest way to identify an alkane is to look at the name of the compound you are dealing with. All alkanes end with -ane. Methane, ethane, propane, butane, pentane, hexane, heptane, octane, nonane, decane, etc. Remembering the structures of the different types is quite simple.

First, remember that the difference in the numbers of carbon and hydrogen atoms is what distinguishes them from each other. If you organize the alkanes in order of ascending numbers of C atoms, you will realize that the difference in C atoms between the alkanes is equal to 1 and the difference in hydrogen atoms is 2.

For example:

Methane= CH_4 (1 C Atom; 4 H Atoms)

Ethane= C_2H_6 (2 C Atoms; 6 H Atoms)

Propane= C_3H_8 (3 C Atoms; 8 H Atoms)

Remembering the number of C atoms per alkane is quite simple, especially for the larger alkanes that practically have the number of C atoms in their name.

*Pent*ane= C_5H_{12}

*Oct*ane= C_8H_{18}

Decane= C10H22

If you struggle to remember the H values, simply memorize the C values, and then double them and add 2 to find them. You can express this as:

H (No. of hydrogen atoms in an alkane)= C (No. of C atoms in an alkane)2 + 2.

This works for all alkanes, even the larger ones like nonadecane (C19H40).

Going back to nomenclature for a moment, "the systematic names for branched hydrocarbons use the lowest possible number to indicate the position of the branch along the longest straight carbon chain in the structure. Thus the systematic name for isobutane is 2-methylpropane, which indicates that a methyl group (a branch consisting of $-CH_3$) is attached to the second carbon of a propane molecule" (Chemistry LibreTexts, 2014).

Alkenes

Let us move on to the alkenes. The names of alkenes work similarly to those of alkanes, making them easy to recognize. Instead of ending their names with -ane, however, alkenes end theirs with -ene (Chemistry LibreTexts, 2014).

Ethylene, also called ethene (C_2H_4 and/or $CH_2=CH_2$) and propylene, or propene (C_3H_6 and/or $CH_3CH=CH_2$) are two of the simplest alkenes. As you can see, both of them, as well as all other alkenes, have at least 1 double C=C bond. Because of this, they have fewer hydrogen atoms than alkanes do (Abozenadah et al., 2017b).

As we have mentioned, the naming of these compounds is very similar to the alkanes. The numbering of C atoms works the same in the nomenclature, but to find the number of H atoms in an alkene, we only have to double the number of C atoms.

For example:

Ethylene= C2H4 (2 C Atoms, 4 H Atoms)

1-Octene= C8H16 (8 C Atoms, 16 H Atoms)

We can express this in the following equation to help *you* remember. DO NOT use this in a test to explain things or as part of an equation.

H (no. of hydrogen atoms in an alkene)= 2(no. of C atoms in the alkene)

Just like alkanes, the boiling point of the alkenes increases as molar mass increases, or as the number of C atoms increases. "For molecules with the same number of carbon atoms and the same general shape, the boiling points usually differ only slightly, just as we would expect for substances whose molar mass differs by only 2 u (equivalent to two hydrogen atoms). Like other hydrocarbons, the alkenes are insoluble in water but soluble in organic solvents" (Abozenadah et al., 2017b).

While alkenes are vital for the production of man-made chemicals and products (such as plastic), they are also widely found throughout nature.

For example, have you ever wondered what gives fruits and vegetables their bright coloring? The isomeric polyenes ($C_{40}H_{56}$) of lycopene and carotenes are what cause the red, orange, and yellow pigmentation in watermelons, tomatoes, carrots, and other fruits and vegetables. Even some vitamins (Vit. A) are derivatives of carotene (Abozenadah et al., 2017b).

Imagine how much color our planet would lose if it wasn't for the alkenes!

Alkynes

Just like alkenes and alkanes, we can easily identify an alkyne by looking at its name. Alkynes all end with the letters -yne (Chemistry LibreTexts, 2014). A shocker indeed.

We distinguish alkynes from the other hydrocarbons by also noting what kind of C-C bonds they have. Alkynes always have a triple C-C bond ($C\equiv C$). The simplest example of an alkyne is the compound acetylene (C_2H_2, or $HC\equiv CH$). Alkynes have a boiling point that is just a bit higher than that of alkenes. But when mixed with oxygen, acetylene burns at an extremely hot temperature (around 3000°C), and is most commonly used in welding tools and those used to cut metals.

The triple bonds that alkynes have explain many of their properties. "Hybridization due to triple bonds allows the uniqueness of the alkyne structure. This triple bond contributes to the nonpolar bonding strength, linearity, and the acidity of alkynes. Physical properties include: nonpolar due to slight solubility in polar solvents; and insoluble in water. This solubility in water and polar solvents is a characteristic feature of alkenes as well" (Nguyen et al., 2013).

Like alkanes and alkenes, alkynes dissolve in organic solvents. Due to the repulsion of electrons, alkynes release a large amount of energy. This high amount of energy can be attributed to the alkyne molecule's energy content. Alkynes and alkenes are very similar to each other in terms of their physical and chemical properties. They are so similar, in fact, that the International Union of Pure and Applied Chemistry (IUPAC) names for alkynes are very similar to alkenes, only their names end in –yne rather than –ene. For example, the IUPAC name for acetylene is ethyne (ethene for alkenes) (Abozenadah et al., 2017b).

Cyclic Hydrocarbons

In cyclic hydrocarbons, the ends of a hydrocarbon chain are connected to form a ring of covalently bonded carbon atoms. The nomenclature of these hydrocarbons is also fairly simple—they steal the names of alkanes, alkenes and alkynes, and then slap on the prefix cyclo- to indicate cyclic hydrocarbons. Among the simplest cyclic alkanes are cyclopropane (C_3H_6), a powerful anesthetic that is also flammable, and cyclobutane (C_4H_8). Typically, when cyclic alkanes are drawn, a polygon is drawn that has the same number of vertex points as there are carbon atoms in the rings. Within these models, each vertex represents a CH_2 atom (Chemistry LibreTexts, 2014).

Alcohols

Alcohols are another type of organic compound that has a wide-reaching scope of compounds within its nomenclature. They are characterized by one or more hydroxyl (–OH) function groups attached to a carbon atom of a hydrocarbon chain, or alkyl group (Ashenhurst, 2014). These alkyl groups are typically represented by the letter R. One can think of alcohols as mere derivatives of water (H_2O) in which one of the H atoms has been replaced by an alkyl group, R (Wade & The Encyclopaedia Britannica, 2019).

Alcohols are among some of the most well-known and commonly found organic compounds. Like hydrocarbons, alcohols have a wide variety of uses, including that of being an intermediate for the synthesis of other compounds. Two of the most well-known alcohols are ethanol and methanol. Ethanol, for example, is the alcohol found in bevarages. Like hydrocarbons, alcohols have their own subcategories.

We call these primary, secondary, and tertiary alcohols. We distinguish them by looking at which carbon in the alkyl group attaches (or is attached) to the hydroxyl group (Wade & The Encyclopaedia Britannica, 2019).

Some alcohols are liquid at room temperature, and others are solids. Their appearance, however, is almost always going to be diaphanous. As the molecular weight of an alcohol increases, its solubility decreases. Alcohol is, generally speaking, quite soluble. This is due to the fact that water, like alcohol, is a hydrogen-bonding solvent, and as such, the dipole of the hydroxyl group can interact favourably with the hydroxyl group of H2O. For example, ethanol and water are miscible in all proportions (Ashenhurst, 2014). However, as molecular weight increases, the alcohol's boiling point, vapour pressures, densities, and viscosities will all increase (Wade & The Encyclopaedia Britannica, 2019).

It should be noted that not all functional groups containing OH are alcohols. When an OH group is attached to a carbonyl (C=O), the functional group is called a "carboxylic acid". When an OH attaches to an alkene, we call the product an "enol". Outside of these are other functional groups (e.g., hydrates) that we won't be considering here because they really belong in a separate category from alcohols.

Their orbitals are similar to those of water as well. While both water and alcohol can be pictured as having an sp^3 hybridized tetrahedral oxygen atom with nonbonding pairs of electrons occupying two of the four sp^3 hybrid orbitals (Wade & The Encyclopaedia Britannica, 2019), there are some key differences between the two.

"Alkyl groups are generally bulkier than hydrogen atoms, however, so the R−O−H bond angle in alcohols is generally larger than the 104.5° H−O−H bond angle in water. For example, the 108.9° bond angle in methanol shows the effect of the methyl group, which is larger than the hydrogen atom of water" (Wade & The Encyclopaedia Britannica, 2019).

Practice Questions

1. **Question: Write the condensed structural formula for the following hydrocarbon: n-heptane**

   Answers and Explanation:

   "To find the condensed structural formula for each hydrocarbon, we use the prefix to determine the number of carbon atoms in the molecule and whether it is cyclic. From the suffix, determine whether multiple bonds are present.

   Next, identify the position of any multiple bonds from the number(s) in the name and then write the condensed structural formula.

   A. A The prefix hept- tells us that this hydrocarbon has seven carbon atoms, and n- indicates that the carbon atoms form a straight chain. The suffix -ane tells that it is an alkane, with no carbon−carbon double or triple bonds. B The structural formula is

CH3CH2CH2CH2CH2CH2CH3, which can also be written as CH3(CH2)5CH3" (Chemistry LibreTexts, 2014).

2. **Question: Write the condensed structural formula for cyclooctene.**

Answer: "The prefix cyclo- tells us that this hydrocarbon has a ring structure, and oct- indicates that it contains eight carbon atoms, which we can draw as an octagon. The suffix -ene tells us that the compound contains a carbon–carbon double bond, but where in the ring do we place the double bond? Because all eight carbon atoms are identical, it doesn't matter. We can draw the structure of cyclooctene as an octagon with *one side* having a double line" (Chemistry LibreTexts, 2014).

3. **Question: Write the condensed structural formula for the following hydrocarbon: 2-pentene**

"Answer: A The prefix pent- tells us that this hydrocarbon has five carbon atoms, and the suffix -ene indicates that it is an alkene, with a carbon–carbon double bond. B The 2- tells us that the double bond begins on the second carbon of the five-carbon atom chain. The condensed structural formula of the compound is therefore CH3CH=CHCH2CH3" (Chemistry LibreTexts, 2014).

4. **Question: Write the condensed structural formula for the following hydrocarbon: 2-butyne**

Answer: The prefix but- tells us that the compound has a chain of four carbon atoms, and the suffix -yne indicates that it has a carbon–carbon triple bond. The 2- tells us that the triple bond begins on the second carbon of the four-carbon atom chain. So the condensed structural formula for the compound is $CH_3C{\equiv}CCH_3$" (Chemistry LibreTexts, 2014).

Chapter 8: Biochemistry

Ever wonder what your skin is made of? How about the last burger you ate, or your favorite salad? The first things that come to mind are probably water and plant and animal cells. Now, think about what the cells that make up those compounds are made of. How do we deal with problems that occur inside of cells? We call this field of study and problem solving biochemistry, the biology-oriented field of chemistry.

Biochemistry as a science explores the chemical processes within and related to living organisms. You can think of it as the part of chemistry that overlaps with biology, and it is one of the sciences that we regard as being laboratory-based. By using what we know about the chemical world as well as applying certain techniques from it, a biochemist can unravel and solve biological problems (Biochemical Society, 2019).

Biochemistry focuses on what's happening inside our cells at a similar level of analysis to chemistry—the molecular level. A biochemist predominantly studies components like proteins, lipids, and organelles. But it is also the study of how cells communicate with each other, the processes that occur in plants, animals, and microorganisms, as well as the changes they undergo during development and life (Vennesland et al., 2019). For example, how do you grow from being a child into an adult? Or how does your immune system fight illnesses? "Biochemists need to understand how the structure of a molecule relates to its function, allowing them to predict how molecules will interact" (Biochemical Society, 2019).

While biochemistry, at its core, is a sort of combination between biology and chemistry, it also incorporates studies and techniques from a variety of other scientific disciplines, including genetics, microbiology, forensics, plant science, and medicine (Biochemical Society, 2019). Due to this variety, biochemistry is an extremely important part of science. Those who pioneer this field are at the forefront of saving and improving lives around the globe.

Biomolecules

Biological molecules, or biomolecules for short, are the molecular substances that are produced by cells and living organisms. We subcategorize these biomolecules into four distinct groups: lipids (fats), proteins, carbohydrates (not to be confused with hydrocarbons) and nucleic acids. When combined, these four make up a large percentage of a cell's mass.

For the remainder of this chapter, we will be taking a look at these four separately, and how they function in the bodies of humans and animals. But first, let us take a look at the stuff that makes up the biomolecules that we will be learning about.

Carbon-Based Life

So far, we know of no life forms that do not require the existence of carbon. Carbon, as we have learned about in the previous chapter, is one of the most common elements and is the most common in organic life forms. Yes, there are other elements that are required too. But carbon is the foundational element of life on earth, and probably anywhere else where it may exist. Carbon has 4 valence electrons, and can thus form four covalent bonds with four other atoms (Molnar et al., 2019). Hence, the most basic organic carbon molecule is CH_4 (methane), as we touched on in the previous chapter.

The more complex molecules, however, form greater chains and can involve a variety of other molecules outside of just hydrogen. However, all of them contain carbon.

Carbohydrates

Most people have a good idea of what carbohydrates are, conceptually. You either love them or hate them. Many people who diet try their best to stay away from them, especially bodybuilders while prepping for show day. If you do anything else, you likely consume mass amounts of them on a daily basis.

Carbohydrates are macromolecules, and despite their infamous nature in the fitness industry, they are an essential part of our diet. Most plant foods, such as grains and fruit, contain high amounts of carbs. The main purpose of carbohydrates is to provide energy to the body, particularly through the simple sugar glucose (blood sugar). They serve other purposes, but we will look at some of them a little later (Molnar et al., 2019).

"Carbohydrates can be represented by the formula $(CH_2O)n$, where n is the number of carbon atoms in the molecule. In other words, the ratio of carbon to hydrogen to oxygen is 1:2:1 in carbohydrate molecules. Carbohydrates are classified into three subtypes: monosaccharides, disaccharides, and polysaccharides" (Molnar et al., 2019).

Carbs can be further subcategorized into sugars (which are the simplest carbohydrates), and starches (which are a more complex carbohydrate that is made up of multiple sugars) (MedlinePlus, 2018).

Now let us look at the different types of carbohydrates in more detail.

Monosaccharides are simple sugars, the most common of which is glucose ($C_6H_{12}O_6$). Glucose is vital for plants and animals to function, as, for example, it provides red blood cells with the energy they need, and is the major blood sugar that is found freely circulating in the blood of higher animals (The Editors of Encyclopaedia Britannica & Rogers, 2019). Glucose releases energy during cellular respiration and is used to help make adenosine triphosphate (ATP). Plants require CO_2 and water to synthesize glucose during photosynthesis, and like animals, that glucose is then used for the energy requirements of the plant (Molnar et al., 2019).

We can identify monosaccharides by the typical number of 3-6 carbon atoms. They can also be identified by the suffix -ose in their names. Monosaccharides can be subcategorized by the number of carbon atoms in the sugar.

Triose= Three carbon atoms

Pentose= Five carbon atoms

Hexose= Six carbon atoms

Monosaccharides are found as either a linear chain or a ring-shaped molecule. When found in aqueous solutions, they are more commonly found in their ring-shape (Molnar et al., 2019).

"Galactose (part of lactose, or milk sugar) and fructose (found in fruit) are other common monosaccharides. Although glucose, galactose, and fructose all have the same chemical formula ($C_6H_{12}O_6$), they differ structurally and chemically (and are known as isomers) because of differing arrangements of atoms in the carbon chain" (Molnar et al., 2019).

Disaccharides form when two monosaccharides combine by undergoing a dehydration reaction (a reaction where a water molecule is removed). This process results in a covalent bond being formed between two sugar molecules when hydroxyl groups (-OH) from one monosaccharide combine with hydrogen atoms from another monosaccharide, which causes the release of an H_2O molecule.

Polysaccharides are a step up from disaccharides, as they are made up of a long chain of monosaccharides linked by covalent bonds instead of just two. The chain may contain a variety of monosaccharides and may be branched or unbranched. Polysaccharides, due to their large chains, are the largest between mono-, di-, and polysaccharides (Molnar et al., 2019).

Starch (excess glucose that is stored in plants), glycogen (excess glucose that is stored in the muscle and liver cells of animals), cellulose, and chitin (found in the exoskeleton of arthropods and the cell walls of fungi) are examples of polysaccharides.

Cellulose [$(C6H10O5)n$] is arguably the most abundant biopolymer on Earth. It is a polysaccharide that is made up of thousands of glucose molecules (Ophardt & Chemistry LibreTexts, 2013). These form a chain, and make up a lot of what we perceive as organic matter, for example, the walls of plant cells and the wood of trees (Molnar et al., 2019).

Cellulose gets its rigidity and high tensile strength (which is important for plant cells) from the fact that "every other glucose monomer in cellulose is flipped over and packed tightly as extended long chains" (Molnar et al., 2019).

Cellulose passing through our digestive system is called dietary fiber. The differences in acetal linkages (carbons that have two ether oxygens attached are considered acetal) between cellulose and starch is what makes cellulose indigestible for humans, as the glucose-glucose bonds in cellulose are unaffected by our digestive enzymes. Herbivores on the other hand (cows, buffalos, horses etc.) are able to digest grass that has very high cellulose levels due to the presence of symbiotic bacteria in their intestinal tract (Ophardt & Chemistry LibreTexts, 2013). These bacteria possess the necessary enzymes, cellulase, (Molnar et al., 2019) to break cellulose down into glucose monomers that can then be utilized as energy by the animal. No vertebrate can digest cellulose directly (Ophardt & Chemistry LibreTexts, 2013).

Lipids

Lipids are a group of compounds that are all identified by their hydrophobic, non-polar nature. Their non-polarity comes from the fact that they only contain non-polar carbon-hydrogen and carbon-carbon bonds. In cells, energy is stored in the form of lipids, more specifically, fats. They also make up the insulating layer of fat (or blubber) in many mammals, and they are also found in the feathers of many birds, allowing water to roll off their feathers without having it weigh them down and rendering them flightless. They are vital for the production of hormones, important for the make-up of cell membranes, and aid in the absorption of fat-soluble vitamins.

Triglycerides (a type of fat molecule) are made up of glycerol and fatty acids. Glycerol is made up of 3 carbon atoms, 5 hydrogen atoms, and 3 hydroxyl ($-$OH) groups. Fatty acids, on the other hand, are made up of a long chain of hydrocarbons that are connected to an acidic carboxyl group at the end of the chain. In biological systems, they normally hold 2$-$28 carbon atoms, but the numbers vary all the way up to 36 (Soult & Chemistry LibreTexts, 2019).

There are also subcategories of fatty acid double bonds—cis and trans. These types describe many different types of fatty acids. Fats are normally solid at room temperature, and oils are almost always liquid (Ahmed & Ahmed, 2018).

In a triglyceride, a fatty acid is connected to each of the 3 oxygen atoms in the −OH groups of the glycerol molecule with a covalent bond. These fats were given the name triglyceride because they are made up of 3 fatty acids [(CH3(CH2)nCOOH)] and because three water molecules are released during covalent bonding.

Fatty acids themselves can be classed as being saturated or unsaturated. A saturated fatty acid is identified by the presence of single bonds between neighboring carbons in the hydrocarbon chain and no other bond type. We call them saturated fatty acids because they are quite literally "saturated with hydrogen; in other words, the number of hydrogen atoms attached to the carbon skeleton is maximized" (Molnar et al., 2019).

When the hydrocarbon chain contains a double bond, we know that the fatty acid is unsaturated.

Most unsaturated fats are commonly referred to as oils, at room temperature at least. Oils are normally extracted from plants (e.g, sunflower oil, olive oil) and contain unsaturated fatty acids. If there is only 1 double bond in the molecule, it is a monounsaturated fat. However, if more than 1 double bond can be found, it is polyunsaturated (Molnar et al., 2019).

Saturated fats tend to turn solid at room temperature quite fast, which is why if you leave a frying pan unwashed after frying a hamburger or some other food rich in fat, you will find thicker fat residue inside the pan after an hour or more. Some examples of saturated animal fats are those that contain stearic acid and palmitic acid contained in meat.

The double bonds found in unsaturated fats prevent fatty acids from packing tightly, meaning they will stay liquid at room temperature. Consuming unsaturated fats will improve your blood cholesterol levels, but eating too much saturated fat (which is found in many foods) will contribute to a build-up of plaque in the arteries, which can cause cardiac arrest (Molnar et al., 2019).

Proteins

Proteins are often called the building blocks of life due to their presence in all life forms. They are made up of hundreds of thousands of amino acids that form a chain to create a protein. There are about 100 amino acids found in nature, but only 20 of them commonly make up a protein chain that, when viewed at a macro level, bears a close resemblance to beads on a string (MedlinePlus, 2020).

We call the amino acids that make up these long chains "α-amino (alpha amino) acids" (Koshland et al., 2019). The α-amino acids get their name from the fact that the α-carbon atom in the molecule contains an amino group ($-NH_2$) as well as a carboxyl group ($-COOH$).

As the pH of acidic solutions drops below 4, the $-COO$ groups react with the hydrogen ions ($H+$) and become uncharged on the product side (–COOH). Alkaline solutions (pH 9+) lead to the loss of a hydrogen ion, and ammonium groups ($-NH^+_3$) become amino groups ($-NH_2$). Throughout the pH levels 4 to 8, amino acids do not migrate in an electrical field because they are both positively and negatively charged. We call such structures dipolar ions and zwitterions (Koshland et al., 2019).

In alpha-amino acids, a peptide bond links the amino group of one amino acid to the carboxyl group of the next amino acid in line. Peptide structures are typically written with the free amino group (which is referred to as the N terminus of a peptide) at the left and the free carboxyl group (also known as the C terminus of a peptide) at the right (Koshland et al., 2019).

Polypeptides are formed as a result of such linkages. Despite the fact that the words 'polypeptide' and 'protein' are often used interchangeably, polypeptides are technically polymers of amino acids, whereas proteins are polypeptides that have combined together, have a distinctive shape, and have a specific function (Molnar et al., 2019).

Enzymes are a common type of protein that act as biological catalysts for biochemical reactions within an animal, such as during digestion, but they can also rearrange bonds and form new ones. Different enzymes bond and break down different substrates when dealing with digestion (Molnar et al., 2019).

Hormones are chemical signaling molecules, usually proteins or steroids, secreted by certain glands or cells of the endocrine system in order to control or regulate various body functions and physiological processes, such as bodily growth, metabolism, and reproduction. Insulin, for example, is a hormone that regulates blood sugar levels, and is important for other functions such as reducing anxiety (Li et al., 2008), which is why eating a medium or high-protein breakfast is a good idea.

Practice Questions

1. **Fatty acids are esterified into mono-, di-, or triglycerides by attaching to:**

Sterol

Cholesterol

Glycerol

Sodium chloride

Answer: Glycerol

An ester is formed when an alcohol combines with an acid in a reaction to form an organic compound. While fatty acids occur in nature in their free (unesterified) state, they are most often found as esters linked to glycerol, cholesterol, or long-chain aliphatic alcohols and as amides in sphingolipids.

AfterWord

All in all, chemistry is probably more complex than most expect, and yet it is so much more interesting as well. Everything from prototypical models of the SI units to quantum theory and the chemistry of life, we've really seen it all on our journey through this most fascinating science.

We learned about solids, liquids, and gases, how their charges impact the way they react with one another, and how the periodic table has far more to it than meets the eye, with all the periodic trends that dictate the order in which the elements are arranged.

We looked at chemical bonds, what ions are, how covalent bonds form, and what the heck the Aufbau principle is. We also learned about different orbitals, the order in which electrons fill them, and the difference between sigma and pi bonds.

Next, we looked at the variety of chemical equations and how we can identify and solve them, and saw how collision theory affects the world of chemistry in terms of how reactants form products and the rate at which they do so.

We learned about electrolysis and electrochemistry; what electrodes, anodes, and cathodes are, as well as all the laws of thermodynamics and the manner in which they pertain to the realm of chemistry. We also learned about equilibrium and why it's harder to clean your room than make it a mess.

We also stopped by Arrhenius, Brønsted, and Lowry to see how their theories differed and where they agreed when it came to acids and bases, and we also learned about the importance of pH and the differences between acids and bases.

Organic chemistry, the art of the chemistry world, is often described by students as being their favorite part of chemistry. Learning about how so much of what we use and interact with is made of the same stuff—hydrocarbons—is either fascinating or terrifying, perhaps both.

And lastly, we looked at how our bodies work through the lens of biochemistry, where we learned about the three carbon-based legos of life; carbohydrates, lipids and proteins, as well as how they are formed, what we need to sustain life, and where they come from.

If you have enjoyed reading this book, please consider leaving a review.

Glossary

Activation Energy: The lowest amount of energy required to activate atoms or molecules to a state that allows them to undergo chemical transformation or physical transport.

Aliphatic: Relating to or denoting organic compounds in which carbon atoms form open chains (as in the alkanes), not aromatic rings.

Amount: A quantity of something, especially the total of a thing or things in number, size, value, or extent.

Ampere: The SI base unit of electrical current.

Amphoteric: (of a compound, especially a metal oxide or hydroxide) able to react both as a base and as an acid.

Anion: Atom(s) carrying a negative electric charge.

Aqueous: Solutions that have water (H_2O) as the solvent. In chemical equations, the symbol (aq).

Aromatic: Relating to or denoting organic compounds containing a planar unsaturated ring of atoms which is stabilized by an interaction of the bonds forming the ring, e.g. benzene and its derivatives.

Arrhenius Theory: Acids are substances that dissociate into hydrogen ions (H^+) or hydroxide ions (OH^-) when placed in water.

Atom: The smallest particle of a chemical element that can exist.

Atomic Mass: The mass of an atom of a chemical element expressed in atomic mass units. It is approximately equivalent to the number of protons and neutrons in the atom (the mass number) or to the average number, allowing for the relative abundances of different isotopes.

Atomic Radius: One-half the distance between the nuclei of identical atoms that are bonded together.

Aufbau Principle: Electrons are added to orbitals as protons are added to an atom. The electron shell is built up as the lower electron orbitals fill before the higher orbitals. Atoms, ions, and molecules form stable electron configurations as a result.

Avogadro's Number: The number used to calculate the amount of particles (atoms, molecules, etc) in a mole or mol.

Benzene: A colorless, volatile liquid hydrocarbon present in coal tar and petroleum, and used in chemical synthesis. Its use as a solvent has been reduced because of its carcinogenic properties.

Bond: A strong force of attraction holding atoms together in a molecule or crystal, resulting from the sharing or transfer of electrons; join or be joined by a chemical bond.

Candela: The SI unit of luminous intensity. One candela is the luminous intensity, in a given direction, of a source that emits monochromatic radiation of frequency 540 × 1012 Hz and that has a radiant intensity in that direction of 1/683 watt per steradian.

Cation: Atoms that bear a positive electric charge.

Celsius: Of or denoting a scale of temperature on which water freezes at 0° and boils at 100° under standard conditions.

Chemical Bond: The interaction between atoms that explains how molecules, ions, crystals, and other stable species are formed in the everyday world.

Chemical Equations: An expression that states the number and type of atoms present in a molecule of a substance, as well as their interactions using symbols.

Chemical Solution: A blend of two or more substances in proportions that can be varied continuously up until the solubility limit.

Collision Theory: The theory used to predict the rate at which chemicals react, particularly in gaseous states.

Dalton: A unit used in expressing the molecular weight of proteins, equivalent to Atomic Mass Unit (AMU)

Delocalized: (of electrons) shared among more than two atoms in a molecule.

Density: The quantity of mass per unit volume of a substance.

Derived Units: A derived unit is a unit of measurement in the International System of Units (SI) that is derived from one or more of the seven base units.

Diatomic: Consisting of two atoms.

Dipole: A molecule in which a concentration of positive electric charge is separated from a concentration of negative charge.

Dissociate: (with reference to a molecule) Split into separate smaller atoms, ions, or molecules, especially reversibly.

D-Orbital: A region where an electron can be found within a certain degree of probability. Once the principal quantum number n equals 3 or greater, the angular quantum number can equal 2, and it is considered the d-orbital.

Double-Displacement Reaction: A type of reaction in which two reactants exchange ions to form two new compounds. The product of double displacement reactions is typically a precipitate.

Electrical Current: A quantity representing the rate of flow of electric charge, usually measured in amperes.

Electron Affinity: The change in energy (in kJ/mole) of a neutral atom (in the gaseous phase) when an electron is added to the atom to form a negative ion. The neutral atom's likelihood of gaining an electron.

Electron Configuration: The arrangement of electrons according to energy levels around an atomic nucleus.

Electronegativity: Symbolized as χ, how likely an atom of a given chemical element is to attract shared electrons during chemical bonding.

Electrostatic: Relating to stationary electric charges or fields as opposed to electric currents.

Endothermic: (of a reaction or process) Accompanied by or requiring the absorption of heat.

Energy: The property of matter and radiation which is manifested as the capacity to perform work (such as causing motion or the interaction of molecules).

Enthalpy: A thermodynamic quantity equivalent to the total heat content of a system. It is equal to the internal energy of the system plus the product of pressure and volume.

Equilibrium: When the concentrations of the products and reactants of a chemical equation are unchanged over time, or when the forward rate of a chemical reaction is equivalent to the backward rate.

Ether: Organic compound that has an oxygen atom bonded to two alkyl or aryl groups.

Exergonic: (of a metabolic or chemical process) Accompanied by the release of energy.

Exothermic: (of a reaction or process) Accompanied by the release of heat. (of a compound) Formed from its constituent elements with a net release of heat.

Fermion: A subatomic particle, such as a nucleon, which has a half-integral spin and follows the statistical description given by Fermi and Dirac.

Intermolecular Forces: The forces that hold atoms together within a molecule, and forces that exist between molecules.

Internal Energy: In thermodynamics, the property or state function that defines the energy of a substance in the absence of effects due to capillarity and external electric, magnetic, and other fields.

International System of Units (SI): The modern form of the metric system and the world's most widely used system of measurement.

Ion: Any atom or group of atoms that bears one or more positive or negative electrical charges.

Ionic Bond: A type of linkage formed by the electrostatic attraction between oppositely charged ions in a chemical compound. Such a bond forms when the valence (outermost) electrons of one atom are transferred permanently to another atom.

Ionization Energy: The amount of energy required to remove an electron from an isolated atom or molecule.

Isotope: Each of two or more forms of the same element that contain equal numbers of protons but different numbers of neutrons in their nuclei, and hence differ in relative atomic mass but not in chemical properties; in particular, a radioactive form of an element.

Kelvin: The SI base unit of thermodynamic temperature (equivalent in size to the degree Celsius), first introduced as the unit used in the Kelvin scale.

Length: The measurement or extent of something from end to end.

Lewis Electron Dot Structure: Also called electron dot structures, these are diagrams that describe the chemical bonding between atoms in a molecule. They also display the total number of lone pairs present in each of the atoms that constitute the molecule.

Luminous: Relating to light as it is perceived by the eye rather than in terms of its actual energy.

Mass: The quantity of matter that an object contains.

Melting Point: The temperature at which a given solid will melt.

Metallic Bond: The force that holds atoms together in a metallic substance. Such a solid consists of closely packed atoms.

Metallic Character: The set of chemical properties that are associated with the elements classified as metals in the periodic table.

Meter: The SI base unit of length (equivalent to approximately 39.37 inches), first introduced as a unit of length in the metric system.

Methodology: A system of methods used in a particular area of study or activity.

Mole: The SI base unit for the amount of substance.

Net Charge: The sum of the charges on an object. The total is determined after taking into account both positive and negative charges.

Nonpolar Covalent Bond: Nonpolar molecules occur when electrons are shared equally between atoms of a diatomic molecule or when polar bonds in a larger molecule cancel each other out.

Octet Rule: The principle that bonded atoms share their eight outer electrons. This gives the atom a valence shell resembling that of a noble gas. The octet rule is a "rule" that is sometimes broken. However, it applies to carbon, nitrogen, oxygen, the halogens, and most metals, especially the alkali metals and alkaline earths.

Orbital: Each of the actual or potential patterns of electron density that may be formed in an atom or molecule by one or more electrons, and can be represented as a wave function.

Overlap: When two or more atomic orbitals of the same type (s,p,d,f) occupy the same space. This results in the sharing of electrons between the orbitals and the formation of molecular orbitals.

Pauli Exclusion Principle: A principle stating that no two electrons can have the identical quantum mechanical state in the same atom or molecule.

Periodic Table: A table of the chemical elements arranged in order of atomic number, usually in rows, so that elements with similar atomic structures (and hence similar chemical properties) appear in vertical columns.

Periodic Trends: A regular variation the properties of an element with increasing atomic number. A periodic trend is attributed to regular variations in the atomic structure of each element.

Pi Bond: A cohesive interaction between two atoms and a pair of electrons that occupy an orbital located in two regions roughly parallel to the line determined by the two atoms. A pair of atoms may be connected by one or by two pi bonds only if a sigma bond also exists between them.

Polar Covalent Bond: A type of covalent bond in which the electrons forming the bond are unequally distributed. In other words, the electrons spend more time on one side of the bond than the other.

Polarity: The distribution of electric charge around atoms, chemical groups, or molecules.

Polyatomic Ion: An ion that is composed of two or more atoms.

P Orbital: A region where an electron can be found, within a certain degree of probability. P orbitals can hold up to 6 electrons.

Scientific Notation: A method for expressing a given quantity as a number having significant digits necessary for a specified degree of accuracy, multiplied by 10 to the appropriate power.

Second: A sixtieth of a minute of time, which as the SI unit of time is defined in terms of the natural periodicity of the radiation of a caesium-133 atom.

Shorthand Electron Configuration: When the electron configuration of a noble gas is written in the electron configuration for any element as its symbol to stand in for the element's core electrons.

Sigma Bond: A mechanism by which two atoms are held together as a result of the forces operating between them and a pair of electrons regarded as shared by them.

Single-Displacement Reactions: A chemical reaction where one reactant is exchanged for one ion of a second reactant. It is also known as a single-replacement reaction.

S Orbital: A region where electrons can be found, within a certain degree of probability. The S orbital can hold up to two electrons.

SPDF Orbitals: The orbital names s (sharp), p (principal), d (diffuse), and f (fundamental) stand for names given to groups of lines originally noted in the spectra of the alkali metals.

Temperature: The degree or intensity of heat present in a substance or object, especially as expressed according to a comparative scale and shown by a thermometer or perceived by touch.

Thermodynamics: The branch of physical science that deals with the relationships between heat and other forms of energy (such as mechanical, electrical, or chemical energy), and, by extension, all forms of energy.

Time: The indefinite continued progress of existence and events in the past, present, and future regarded as a whole.

Unit: A quantity chosen as a standard in terms of which other quantities may be expressed.

Valence Bond Theory: A chemical bonding theory that explains the chemical bonding between two atoms. It states that bonding is caused by the overlap of half-filled atomic orbitals.

Valence Electron: The outermost electrons in an atom, the most susceptible to participating in chemical bond formation or ionization.

Valence Shell (of an atom): The outermost shell of an atom containing the valence electrons.

Volume: The amount of space that a substance or object occupies, or that is enclosed within a container.

Weight: The force exerted on the mass of a body by a gravitational field.

References

Abozenadah, H., Bishop, A., Bittner, S., Lopez, C., Wiley, C., & Flatt, P. M. (2017a, January 10). *Chapter 1: Measurements in chemistry– chemistry.* Wou.edu. https://wou.edu/chemistry/courses/online-chemistry-textbooks/foundations-general-organic-biological-chemistry/chapter-1-measurements-chemistry/

Abozenadah, H., Bishop, A., Bittner, S., Lopez, O., Wiley, C., & Flatt, P. M. (2017b, January 10). *CH105: Chapter 8– alkenes, alkynes and aromatic compounds– chemistry.* Wou.edu. https://wou.edu/chemistry/courses/online-chemistry-textbooks/ch105-consumer-chemistry/ch105-chapter-8/

Agnew, M. A., Agnew, H., & LibreTexts. (2019, August 22). 9.2: *Aliphatic hydrocarbons.* Chemistry LibreTexts. https://chem.libretexts.org/Courses/Sacramento_City_College/SCC%3A_CHEM_330_-_Adventures_in_Chemistry_(Alviar-Agnew)/09%3A_Organic_Chemistry/9.02%3A_Aliphatic_Hydrocarbons#:~:text=Aliphatic%20hydrocarbons%20are%20hydrocarbons%20based

Ahmed, S., & Ahmed, O. (2018, October 27). *Biochemistry, lipids.* Nih.gov; StatPearls Publishing. https://www.ncbi.nlm.nih.gov/books/NBK525952/

American Chemical Society. (n.d.). *Organic chemistry.* American Chemical Society. https://www.acs.org/content/acs/en/careers/chemical-sciences/areas/organic-chemistry.html#:~:text=Organic%20chemistry%20is%20the%20study

AP Chemistry Varsity Tutors. (n.d.). *Kinetics and energy - AP chemistry.* Www.varsitytutors.com. Retrieved May 20, 2022, from https://www.varsitytutors.com/ap_chemistry-help/thermochemistry-and-kinetics/kinetics-and-energy?page=2

Ashenhurst, J. (2014, September 17). *Alcohols (1) - nomenclature and properties. Master Organic Chemistry.* https://www.masterorganicchemistry.com/2014/09/17/alcohols-1-nomenclature-and-properties/

Augustyn, A., & Encyclopaedia Britannica. (2021). *Collision theory.* In Encyclopædia Britannica. https://www.britannica.com/science/collision-theory-chemistry

Averill, B. (2013, November 26). *20.4: The alkali metals (group 1)*.
    Libretexts.
    https://chem.libretexts.org/Bookshelves/General_Chemistry/
    Book%3A_General_Chemistry%3A_Principles_Patterns_and_
    Applications_(Averill)/20%3A_Periodic_Trends_and_the_s-
    Block_Elements/20.04%3A_The_Alkali_Metals_(Group_1)

Ball, D. W., & Key, J. A. (2014, September 16). *Oxidation-Reduction
    reactions*.   Opentextbc.ca.
    https://opentextbc.ca/introductorychemistry/chapter/chapter-
    4-oxidation-reduction-reactions/

Bertrand, G., Guess, J., & Johns, D. (2013, October 2). *Dipole-Dipole
    interactions*.   Chemistry   LibreTexts.
    https://chem.libretexts.org/Bookshelves/Physical_and_Theore
    tical_Chemistry_Textbook_Maps/Supplemental_Modules_(Ph
    ysical_and_Theoretical_Chemistry)/Physical_Properties_of_
    Matter/Atomic_and_Molecular_Properties/Intermolecular_Fo
    rces/Specific_Interactions/Dipole-Dipole_Interactions

Betts, J. (2016, July 27). *Examples of ways to measure volume*.
    YourDictionary.
    https://examples.yourdictionary.com/examples-of-ways-to-
    measure-volume.html

Biochemical Society. (2019). *What is biochemistry? -*. Biochemistry.org.
    https://biochemistry.org/education/careers/becoming-a-
    bioscientist/what-is-biochemistry/

Broderick, C., Moussa, M., Clark, J., & Chemistry LibreTexts. (2019,
    February 23). *Overview of acids and bases*. Chemistry
    LibreTexts.
    https://chem.libretexts.org/Bookshelves/Physical_and_Theore
    tical_Chemistry_Textbook_Maps/Supplemental_Modules_(Ph
    ysical_and_Theoretical_Chemistry)/Acids_and_Bases/Acid/O
    verview_of_Acids_and_Bases

Butane.Chem. (n.d.). *Worksheet 25- oxidation/reduction reactions
    oxidation number rules (p. 1)*. Retrieved May 20, 2022, from
    http://butane.chem.uiuc.edu/anicely/chem102Dfa10/Workshe
    ets/Worksheet25_Redox_Key.pdf

Byju's. (n.d.). *Thermodynamics questions - practice questions of
    thermodynamics with answer & explanations*. Byjus.com.
    Retrieved   May   30,   2022,   from
    https://byjus.com/chemistry/thermodynamics-questions/

Carey, F. A., & The Encyclopaedia Britannica. (2018). *Hydrocarbon |
    definition, types, & facts*. In Encyclopædia Britannica.
    https://www.britannica.com/science/hydrocarbon

Chemistry LibreTexts. (2013, October 2). *Periodic trends*. Chemistry
    LibreTexts.
    https://chem.libretexts.org/Bookshelves/Inorganic_Chemistry

/Supplemental_Modules_and_Websites_(Inorganic_Chemistr
y)/Descriptive_Chemistry/Periodic_Trends_of_Elemental_Pr
operties/Periodic_Trends#:~:text=Major%20periodic%20tren
ds%20include%3A%20electronegativity

Chemistry LibreTexts. (2014, June 19). *3.7: Names of formulas of organic compounds*. Chemistry LibreTexts. https://chem.libretexts.org/Bookshelves/General_Chemistry/ Map%3A_General_Chemistry_(Petrucci_et_al.)/03%3A_Che mical_Compounds/3.7%3A__Names_of_Formulas_of_Organi c_Compounds

Chemistry LibreTexts. (2016a, June 27). *9.7: Multiple covalent bonds*. Chem.LibreTexts.org. https://chem.libretexts.org/Bookshelves/Introductory_Chemis try/Introductory_Chemistry_(CK-12)/09%3A_Covalent_Bonding/9.07%3A_Multiple_Covalent_ Bonds

Chemistry LibreTexts. (2016b, December 9). *4.4: Polar and non-polar covalent bonds*. Chemistry LibreTexts. https://chem.libretexts.org/Courses/Eastern_Mennonite_Univ ersity/EMU%3A_Chemistry_for_the_Life_Sciences_(Cessna)/ 4%3A_Covalent_Bonding_and_Simple_Molecular_Compound s/4.4%3A_Polar_and_Non-polar_Covalent_Bonds

Chemistry LibreTexts. (2017, August 29). *6.8: Electron configurations*. Chemistry LibreTexts. https://chem.libretexts.org/Courses/University_of_Missouri/ MU%3A__1330H_(Keller)/06._Electronic_Structure_of_Ato ms/6.8%3A_Electron_Configurations

Chemistry LibreTexts. (2019a, June 3). *Aufbau principle*. Chemistry LibreTexts. https://chem.libretexts.org/Bookshelves/Physical_and_Theore tical_Chemistry_Textbook_Maps/Supplemental_Modules_(Ph ysical_and_Theoretical_Chemistry)/Electronic_Structure_of_ Atoms_and_Molecules/Electronic_Configurations/Aufbau_Pri nciple

Chemistry LibreTexts. (2019b, August 7). *Electron affinity*. Chemistry LibreTexts. https://chem.libretexts.org/Bookshelves/Physical_and_Theore tical_Chemistry_Textbook_Maps/Supplemental_Modules_(Ph ysical_and_Theoretical_Chemistry)/Physical_Properties_of_ Matter/Atomic_and_Molecular_Properties/Electron_Affinity

Chemistry LibreTexts. (2021, January 10). *6.E: Electrochemistry (practice problems with answers)*. Chemistry LibreTexts. https://chem.libretexts.org/Courses/Prince_Georges_Commu nity_College/CHEM_1020%3A_General_Chemistry_II_(S.N.

_Yasapala)/06%3A_Electrochemistry/6.E%3A_Electrochemist ry_(Practice_problems_with_Answers)

Chemistry LibreTexts. (2022, April 30). *11.5: Decomposition reactions.* Chemistry LibreTexts. https://chem.libretexts.org/Bookshelves/Introductory_Chemis try/Introductory_Chemistry_(CK-12)/11%3A_Chemical_Reactions/11.05%3A_Decomposition_R eactions

Chemistry Libretexts. (2020, August 16). *Le chatelier's principle.* Chemistry LibreTexts. https://chem.libretexts.org/Bookshelves/Physical_and_Theore tical_Chemistry_Textbook_Maps/Supplemental_Modules_(Ph ysical_and_Theoretical_Chemistry)/Equilibria/Le_Chateliers_ Principle

Chemistry LibreTexts & Editors. (2016, June 27). *9.18: Sigma and pi bonds.* Chemistry LibreTexts. https://chem.libretexts.org/Bookshelves/Introductory_Chemis try/Introductory_Chemistry_(CK-12)/09%3A_Covalent_Bonding/9.18%3A_Sigma_and_Pi_Bon ds

Chemistry LibreTexts, & CK-12. (2016a, June 27). *21.12: Strong and weak acids and acid ionization constant.* Chemistry LibreTexts. https://chem.libretexts.org/Bookshelves/Introductory_Chemis try/Introductory_Chemistry_(CK-12)/21%3A_Acids_and_Bases/21.12%3A_Strong_and_Weak_ Acids_and_Acid_Ionization_Constant_(K_texta)

Chemistry LibreTexts, & CK-12. (2016b, June 27). *21.17: Titration experiment.* Chemistry LibreTexts. https://chem.libretexts.org/Bookshelves/Introductory_Chemis try/Introductory_Chemistry_(CK-12)/21%3A_Acids_and_Bases/21.17%3A_Titration_Experimen t

Chemistry Varsity Tutors. (n.d.). *Help with Single-Replacement Reactions - High School Chemistry.* Www.varsitytutors.com. Retrieved May 19, 2022, from https://www.varsitytutors.com/high_school_chemistry-help/help-with-single-replacement-reactions

ChemistryStudy.com. (2022). *How to identify a double displacement reaction.* Study.com. https://study.com/skill/learn/how-to-identify-a-double-displacement-reaction-explanation.html

Chieh, C. (2015, August 28). *Electrolysis.* Chemistry LibreTexts. https://chem.libretexts.org/Bookshelves/Analytical_Chemistry /Supplemental_Modules_(Analytical_Chemistry)/Electrochem istry/Basics_of_Electrochemistry/Electrochemistry/Electrolysi s

Chieh, C., & Chemistry LibreTexts. (2015, September 7). *Electrochemistry review*. Chemistry LibreTexts. https://chem.libretexts.org/Bookshelves/Analytical_Chemistry /Supplemental_Modules_(Analytical_Chemistry)/Electrochem istry/Basics_of_Electrochemistry/Electrochemistry/Electroche mistry_Review

Chieh, C., & LibreTexts. (2015, August 26). *Electrochemistry*. Chemistry LibreTexts. https://chem.libretexts.org/Bookshelves/Analytical_Chemistry /Supplemental_Modules_(Analytical_Chemistry)/Electrochem istry/Basics_of_Electrochemistry/Electrochemistry

Clark, J. (2019, February 23). *Metallic bonding*. Chemistry LibreTexts. https://chem.libretexts.org/Bookshelves/Physical_and_Theore tical_Chemistry_Textbook_Maps/Supplemental_Modules_(Ph ysical_and_Theoretical_Chemistry)/Chemical_Bonding/Fund amentals_of_Chemical_Bonding/Metallic_Bonding

College Chemistry Varsity Tutors. (n.d.-a). *Combustion reactions - college chemistry*. Www.varsitytutors.com. Retrieved May 20, 2022, from https://www.varsitytutors.com/college_chemistry-help/combustion-reactions

College Chemistry Varsity Tutors. (n.d.-b). *Redox reactions - college chemistry*. Www.varsitytutors.com. Retrieved May 21, 2022, from https://www.varsitytutors.com/college_chemistry-help/reactions/redox-reactions

College Varsity Tutors. (n.d.). *Thermodynamics - college chemistry*. Www.varsitytutors.com. Retrieved May 30, 2022, from https://www.varsitytutors.com/college_chemistry-help/thermodynamics-and-kinetics/thermodynamics

Cotton, F. A. (2019). Transition metal | *Definition, properties, elements, & facts*. In Encyclopædia Britannica. https://www.britannica.com/science/transition-metal

CrashCourse. (2013). Equilibrium: Crash course chemistry #28. In YouTube. https://www.youtube.com/watch?v=g5wNg_dKsYY

CrashCourse, & Green, H. (2013). *The periodic table: Crash course chemistry #4*. In YouTube. https://www.youtube.com/watch?v=0RRVV4Diomg

DeWitt, T. (2015a). *Introduction to oxidation reduction (redox) reactions [YouTube Video]*. In YouTube. https://www.youtube.com/watch?v=5rtJdjas-mY

DeWitt, T. (2015b). *Electrolysis*. In YouTube. https://www.youtube.com/watch?v=dRtSjJCKkIo

Faizi, S., & Chemistry LibreTexts. (2016, May 13). *2.4 electron configurations*. Chemistry LibreTexts. https://chem.libretexts.org/Courses/Valley_City_State_Univer

sity/Chem_115/Chapter_2%3A_Atomic_Structure/2.4_Electro
n_Configurations

Flowers, P., Robinson, PhD, W., Langley, R., & Theopold, K. (2015, March 11). *2.1 Early ideas in atomic theory - chemistry.* Openstax.org. https://openstax.org/books/chemistry/pages/2-1-early-ideas-in-atomic-theory

Flowers, P., Theopold, K., Langley, R., Clark, A., OpenStax College, & Crash Course Chemistry. (2018, November 26). *3.1: Electron configurations.* Chemistry LibreTexts. https://chem.libretexts.org/Courses/Oregon_Institute_of_Tec hnology/OIT%3A_CHE_202_-_General_Chemistry_II/Unit_3%3A_Periodic_Patterns/3.1%3 A_Electron_Configurations

GRE Chemistry Varsity Tutors. (n.d.). *General thermodynamics - GRE subject test: Chemistry.* Www.varsitytutors.com. Retrieved May 27, 2022, from https://www.varsitytutors.com/gre_subject_test_chemistry-help/thermodynamics-and-phases/physical-chemistry/general-thermodynamics

Harper College. (2019). *Qualitative kinetics - kinetics and collision theory.* Harpercollege.edu. http://dept.harpercollege.edu/chemistry/chm/100/dgodambe/ thedisk/kinetic/6back.htm

Harvey, D., & Chemistry LibreTexts. (2019, January 10). 6.2: Thermodynamics and equilibrium chemistry. Chemistry LibreTexts. https://chem.libretexts.org/Bookshelves/Analytical_Chemistry/Analyti cal_Chemistry_2.1_(Harvey)/06%3A_Equilibrium_Chemistry/6.02%3 A_Thermodynamics_and_Equilibrium_Chemistry

Helmenstine, A. M. (2011). What is a double displacement reaction? ThoughtCo. https://www.thoughtco.com/definition-of-double-displacement-reaction-605045

Helmenstine, A. M. (2019a). What is a single-displacement reaction? Definition and examples. ThoughtCo. https://www.thoughtco.com/definition-of-single-displacement-reaction-605662

Helmenstine, A. M. (2019b, May 6). What are the three laws of thermodynamics? ThoughtCo. https://www.thoughtco.com/laws-of-thermodynamics-p3-2699420#:~:text=The%20Third%20Law%20of%20Thermodynamics&t ext=It%20is%20impossible%20to%20reduce

Helmenstine, A. M. (2019c, July 3). How to define a pi bond in chemistry. ThoughtCo. https://www.thoughtco.com/definition-of-pi-bond-605519

Helmenstine, A. M. (2020a, August 25). What do the numbers on the periodic table mean? ThoughtCo. https://www.thoughtco.com/the-

numbers-on-the-periodic-table-
608806#:~:text=Most%20periodic%20tables%20include%20a

Helmenstine, A. M. (2020b, August 28). What are good examples of chemical energy? ThoughtCo. https://www.thoughtco.com/example-of-chemical-energy-
609260#:~:text=Chemical%20batteries%3A%20Store%20chemical%2
0energy

Helmenstine, A. M. (2020c, August 28). What do s, p, d, and f mean in chemistry, and why do you need them? ThoughtCo. https://www.thoughtco.com/angular-momentum-quantum-numbers-
606461#:~:text=The%20orbital%20names%20s%2C%20p

Helmenstine, A. M. (2020d, August 28). What is ionization energy? ThoughtCo. https://www.thoughtco.com/ionization-energy-and-trend-
604538#:~:text=Ionization%20energy%20exhibits%20periodicity%20
on

Helmenstine, A. M. (2020e, August 28). What the aufbau principle means in chemistry. ThoughtCo. https://www.thoughtco.com/definition-of-aufbau-principle-604805

Helmenstine, A. M. (2020f, September 2). Examples of polar and nonpolar molecules. ThoughtCo. https://www.thoughtco.com/examples-of-polar-and-nonpolar-molecules-608516

Helmenstine, A. M. (2020g, December 2). Easy steps to balance chemical equations. ThoughtCo. https://www.thoughtco.com/how-to-balance-chemical-equations-603860

Helmenstine, A. M. (2021a, February 16). Know the difference between ionic and covalent bonds. ThoughtCo. https://www.thoughtco.com/ionic-and-covalent-chemical-bond-differences-606097#:~:text=Key%20Points

Helmenstine, A. M. (2021b, April 1). Definition and examples of a polar bond in chemistry. ThoughtCo. https://www.thoughtco.com/definition-of-polar-bond-and-examples-605530

Helmenstine, A. M., & ThoughtCo. (2019, September 12). Understanding endothermic and exothermic reactions. ThoughtCo. https://www.thoughtco.com/endothermic-and-exothermic-reactions-602105

Helmenstine, A. M., & ThoughtCo. (2020a, January 13). Combustion: Definition and equation. ThoughtCo. https://www.thoughtco.com/definition-of-combustion-
605841#:~:text=Combustion%20is%20a%20chemical%20reaction%20
that%20occurs%20between%20a%20fuel

Helmenstine, A. M., & ThoughtCo. (2020b, August 28). Here's how metallic bonding works. ThoughtCo.com. https://www.thoughtco.com/metallic-bond-definition-properties-and-examples-4117948

Jha, S. K. (2020, May 7). *Non-benzenoid aromatic compounds.* MLTCollege.org. https://mltcollege.org/wp-content/uploads/2020/05/Non-benzenoid-aromatic-compounds.pdf

Koshland, D. E., Haurowitz, F., & The Encyclopaedia Britannica. (2019). *Protein | definition, structure, & classification.* In Encyclopædia Britannica. https://www.britannica.com/science/protein

Kramer, M. (2017, April 24). *How the mass of an object affects its motion.* Sciencing. https://sciencing.com/mass-object-affects-its-motion-10044594.html

Lambert, F. L., & LibreTexts. (2018, January 23). *Microstates.* Chemistry LibreTexts. https://chem.libretexts.org/Bookshelves/Physical_and_Theoretical_Chemistry_Textbook_Maps/Supplemental_Modules_(Physical_and_Theoretical_Chemistry)/Thermodynamics/Energies_and_Potentials/Entropy/Microstates#:~:text=Each%20specific%20way%2C%20each%20arrangement

Lawson, P., Lower, S., & Chemistry LibreTexts. (2020, August 22). *The collision theory.* Chemistry LibreTexts. https://chem.libretexts.org/Bookshelves/Physical_and_Theoretical_Chemistry_Textbook_Maps/Supplemental_Modules_(Physical_and_Theoretical_Chemistry)/Kinetics/06%3A_Modeling_Reaction_Kinetics/6.01%3A_Collision_Theory/6.1.06%3A_The_Collision_Theory

Li, C., Barker, L., Ford, E. S., Zhang, X., Strine, T. W., & Mokdad, A. H. (2008). *Diabetes and anxiety in US adults: Findings from the 2006 behavioral risk factor surveillance system.* Diabetic Medicine, 25(7), 878–881. https://doi.org/10.1111/j.1464-5491.2008.02477.x

LibreText. (2016, June 27). *9.6: Single covalent bonds.* Chemistry LibreTexts. https://chem.libretexts.org/Bookshelves/Introductory_Chemistry/Introductory_Chemistry_(CK-12)/09%3A_Covalent_Bonding/9.06%3A_Single_Covalent_Bonds

LibreTexts. (2014a, June 27). *Temperature basics.* Chemistry LibreTexts. https://chem.libretexts.org/Bookshelves/Analytical_Chemistry/Supplemental_Modules_(Analytical_Chemistry)/Quantifying_Nature/Temperature_Basics

LibreTexts. (2014b, November 20). *Chapter 10.1: Gaseous elements and compounds.* Chem.libretexts. https://chem.libretexts.org/Courses/Howard_University/General_Chemistry%3A_An_Atoms_First_Approach/Unit_4%3A_

_Thermochemistry/Chapter_10%3A_Gases/Chapter_10.1%3A
_Gaseous_Elements_and_Compounds

LibreTexts. (2019, February 23). *Ionic and covalent bonds.* Chemistry LibreTexts.
https://chem.libretexts.org/Bookshelves/Organic_Chemistry/S
upplemental_Modules_(Organic_Chemistry)/Fundamentals/I
onic_and_Covalent_Bonds

LibreTexts. (2020, May 28). *2.1.1: Practice problems- A history of atomic theory (Optional).* Chemistry LibreTexts.
https://chem.libretexts.org/Courses/Oregon_Tech_PortlandM
etro_Campus/OT_-_PDX_-
_Metro%3A_General_Chemistry_I/02%3A_Atoms_and_Elem
ents/2.01%3A_A_History_of_Atomic_Theory/2.1.01%3A_Pra
ctice_Problems-_A_History_of_Atomic_Theory_(Optional)

LibreTexts, & Chieh, C. (2015, August 26). *Galvanic cells.* Chemistry LibreTexts.
https://chem.libretexts.org/Bookshelves/Analytical_Chemistry
/Supplemental_Modules_(Analytical_Chemistry)/Electrochem
istry/Basics_of_Electrochemistry/Electrochemistry/Galvanic_
Cells

LibreTexts, & Soult Ph.D, A. (2019, May 3). *5.3: Types of chemical reactions.* Chemistry LibreTexts.
https://chem.libretexts.org/Courses/Valley_City_State_Univer
sity/Chem_121/Chapter_5%3A_Introduction_to_Redox_Che
mistry/5.3%3A_Types_of_Chemical_Reactions

Ling, S. J., Moebs, W., & Sanny, J. (2016, September 15). *Reversible and irreversible processes.* Opentextbc.ca; OpenStax.
https://opentextbc.ca/universityphysicsv2openstax/chapter/re
versible-and-irreversible-processes/

Malley, K., Singh, R., Duan, T., & Chemistry LibreTexts. (2019, September 30). *2nd law of thermodynamics.* Chemistry LibreTexts.
https://chem.libretexts.org/Bookshelves/Physical_and_Theore
tical_Chemistry_Textbook_Maps/Supplemental_Modules_(Ph
ysical_and_Theoretical_Chemistry)/Thermodynamics/The_Fo
ur_Laws_of_Thermodynamics/Second_Law_of_Thermodyna
mics

MedlinePlus. (2018, June 18). *Carbohydrates.* Medlineplus.gov.
https://medlineplus.gov/carbohydrates.html#:~:text=Carbohy
drates%2C%20or%20carbs%2C%20are%20sugar

MedlinePlus. (2020, September 18). *What are proteins and what do they do?: MedlinePlus genetics.* Medlineplus.gov.
https://medlineplus.gov/genetics/understanding/howgeneswo
rk/protein/#:~:text=Proteins%20are%20large%2C%20comple
x%20molecules

Mersha, A., & Chemistry LibreTexts. (2013, October 2). *3rd law of thermodynamics.* Chemistry LibreTexts. https://chem.libretexts.org/Bookshelves/Physical_and_Theore tical_Chemistry_Textbook_Maps/Supplemental_Modules_(Ph ysical_and_Theoretical_Chemistry)/Thermodynamics/The_Fo ur_Laws_of_Thermodynamics/Third_Law_of_Thermodynami cs#:~:text=The%203rd%20law%20of%20thermodynamics

Referencing Atkins, Peter and de Paula. *Physical Chemistry for the Life Sciences*, Oxford University Press. Oxford, UK. 2006.

Molnar, C., Gair, J., & OpentTextBC. (2019). *2.3 biological molecules – concepts of biology-1st canadian edition.* Opentextbc.ca. https://opentextbc.ca/biology/chapter/2-3-biological-molecules/

NASA Global Climate Change. (2016, September 12). *10 interesting things about air.* Climate Change: Vital Signs of the Planet. https://climate.nasa.gov/news/2491/10-interesting-things-about-air/#:~:text=It

Nguyen, B. K., Chin, G. M., & Chemistry LibreTexts. (2013, October 2). *Properties and bonding in the alkynes.* Chemistry LibreTexts. https://chem.libretexts.org/Bookshelves/Organic_Chemistry/S upplemental_Modules_(Organic_Chemistry)/Alkynes/Properti es_of_Alkynes/Properties_and_Bonding_in_the_Alkynes

OpenStax. (2015, September 29). *2.1 Atoms, isotopes, ions, and molecules: The building blocks - biology.* Openstax.org; OpenStax, Biology. https://cnx.org/contents/GFy_h8cu@9.87:vogY0C26@12/Ato ms-Isotopes-Ions-and-Molecules-The-Building-Blocks

OpenStax. (2016). *8.2 hybrid atomic orbitals.* Opentextbc.ca; Pressbooks. https://opentextbc.ca/chemistry/chapter/8-2-hybrid-atomic-orbitals/

OpenTextBC. (2019). *10.6 lattice structures in crystalline solids– Chemistry.* Opentextbc.ca. https://opentextbc.ca/chemistry/chapter/10-6-lattice-structures-in-crystalline-solids/

Ophardt, C., & Chemistry LibreTexts. (2013, October 2). *Cellulose.* Chemistry LibreTexts. https://chem.libretexts.org/Bookshelves/Biological_Chemistry /Supplemental_Modules_(Biological_Chemistry)/Carbohydrat es/Polysaccharides/Cellulose

Percy Bell, R., & Encyclopaedia Britannica. (2018). *Acid-base reaction | definition, examples, formulas, & facts.* In Encyclopædia Britannica. https://www.britannica.com/science/acid-base-reaction

PSIBERG Team. (2022, January 3). *Zeroth law of thermodynamics: The thermal equilibrium law.* PSIBERG. https://psiberg.com/zeroth-law-of-thermodynamics/

Pulido, K., Chappell, C., & McDuff, A. (2013, October 2). *Effect of temperature on equilibrium.* Chemistry LibreTexts. https://chem.libretexts.org/Bookshelves/Physical_and_Theoretical_Chemistry_Textbook_Maps/Supplemental_Modules_(Physical_and_Theoretical_Chemistry)/Equilibria/Le_Chateliers_Principle/Effect_Of_Temperature_On_Equilibrium_Composition

Purdue University Department of Chemistry. (n.d.). *Dipole-Dipole forces.* Purdue.edu. Retrieved May 7, 2022, from https://www.chem.purdue.edu/gchelp/liquids/dipdip.html

Robert, J. D., Caserio, M. C., & LibreTexts. (2014, November 26). *3.6: Arenes.* Chemistry LibreTexts. https://chem.libretexts.org/Bookshelves/Organic_Chemistry/Book%3A_Basic_Principles_of_Organic_Chemistry_(Roberts_and_Caserio)/03%3A_Organic_Nomenclature/3.06%3A_Arenes

ScienceABC. (2020, October 29). *Octet rule: Why are atoms with 8 valence electrons so stable?* (A. Thombre, Ed.). Science ABC. https://www.scienceabc.com/pure-sciences/why-are-atoms-with-8-valence-electrons-so-stable.html

Siegel, E. (2020, April 16). *You are not mostly empty space.* Forbes. https://www.forbes.com/sites/startswithabang/2020/04/16/you-are-not-mostly-empty-space/?sh=7d0036e12c2b

Soult, A., & Chemistry LibreTexts. (2019, June 5). *14.2: Lipids and triglycerides.* Chemistry LibreTexts. https://chem.libretexts.org/Courses/University_of_Kentucky/UK%3A_CHE_103_-_Chemistry_for_Allied_Health_(Soult)/Chapters/Chapter_14%3A_Biological_Molecules/14.2%3A_Lipids_and_Triglycerides

The Editors of Encyclopaedia Britannica. (1998a). *Metalloid | chemistry.* In Encyclopædia Britannica. https://www.britannica.com/science/metalloid

The Editors of Encyclopaedia Britannica. (1998b, July 20). *Electrode | electronics.* Encyclopedia Britannica. https://www.britannica.com/science/electrode

The Editors of Encyclopaedia Britannica. (2019a). *Atomic mass | definition & facts.* In Encyclopædia Britannica. https://www.britannica.com/science/atomic-mass

The Editors of Encyclopaedia Britannica. (2019b). *Mole | definition, number, & facts.* In Encyclopædia Britannica. https://www.britannica.com/science/mole-chemistry

The Editors of Encyclopaedia Britannica. (2021). *Charles-Augustin de coulomb | french physicist.* In Encyclopædia Britannica. https://www.britannica.com/biography/Charles-Augustin-de-Coulomb

The Editors of Encyclopaedia Britannica, & Gregersen, E. (2016). *Coulomb force | physics | Britannica.* In E. Rodriguez (Ed.), Encyclopædia Britannica. https://www.britannica.com/science/Coulomb-force

The Editors of Encyclopaedia Britannica, & Gregersen, E. (2019). *Law of multiple proportions | chemistry.* In Encyclopædia Britannica. https://www.britannica.com/science/law-of-multiple-proportions

The Editors of Encyclopaedia Britannica, & Hosch, W. (2019). *Atomic theory | physics | Britannica.* Encyclopædia Britannica. https://www.britannica.com/science/atomic-theory

The Editors of Encyclopaedia Britannica, & L. Hosch, W. (2009). *Coulomb's law | definition & facts | britannica.* In Encyclopædia Britannica. https://www.britannica.com/science/Coulombs-law

The Editors of Encyclopaedia Britannica, & L. Hosch, W. (2019). *Metallic bond | chemistry.* In Encyclopædia Britannica. https://www.britannica.com/science/metallic-bond

The Editors of Encyclopaedia Britannica, & Rodriguez, E. (2019). *Nonmetal | chemistry.* In Encyclopædia Britannica. https://www.britannica.com/science/nonmetal

The Editors of Encyclopaedia Britannica, & Rogers, K. (2019). *Glucose | definition, structure, & function.* In Encyclopædia Britannica. https://www.britannica.com/science/glucose

The Editors of Encyclopedia Britannica. (2011). *Electronegativity | physics.* In Encyclopædia Britannica. https://www.britannica.com/science/electronegativity

The Editors of Encyclopedia Britannica. (2014). *Ionic bond | chemistry.* In Encyclopædia Britannica. https://www.britannica.com/science/ionic-bond

The Editors of Encyclopedia Britannica. (2016). *Arrhenius theory | chemistry.* In Encyclopædia Britannica. https://www.britannica.com/science/Arrhenius-theory

The Editors of Encyclopedia Britannica, & Augustyn, A. (2017). *Thomson atomic model.* In Encyclopædia Britannica. https://www.britannica.com/science/Thomson-atomic-model

The Editors of Encyclopedia Britannica, Augustyn, A., & Schreiber, B. (2019). *Metal | definition, characteristics, types, & facts.* In Encyclopædia Britannica. https://www.britannica.com/science/metal-chemistry

The Editors of the Encyclopedia Britannica. (2016). *Chemical equilibrium*. In Encyclopædia Britannica. https://www.britannica.com/science/chemical-equilibrium

The Encyclopaedia Britannica, & Gregersen, E. (2022, May 12). *Lye | chemical compound | britannica*. Www.britannica.com. https://www.britannica.com/science/lye

The Organic Chemistry Tutor. (2017). *First law of thermodynamics, basic introduction - internal energy, heat and work - chemistry*. In YouTube. https://www.youtube.com/watch?v=NyOYWo7-L5g

Thomas, A. W., & W. Weise. (2001). *The structure of the nucleon*. Wiley-Vch.

U.S Department of the Interior, & Water Science School. (2019, October 22). *PH and water | U.S geological survey*. Www.usgs.gov. https://www.usgs.gov/special-topics/water-science-school/science/ph-and-water#overview

Urone, P. P., Hinrichs, R., & OpenStax. (2012). *The first law of thermodynamics - college physics - openstax*. Openstax.org. https://openstax.org/books/college-physics/pages/15-1-the-first-law-of-thermodynamics

Varsity Tutors. (n.d.-a). *Electron affinity - GRE subject test: Chemistry*. Www.varsitytutors.com. Retrieved May 3, 2022, from https://www.varsitytutors.com/gre_subject_test_chemistry-help/electron-affinity

Varsity Tutors. (n.d.-b). *Ionic bonds - MCAT physical*. Www.varsitytutors.com. Retrieved May 11, 2022, from https://www.varsitytutors.com/mcat_physical-help/ionic-bonds

Varsity Tutors. (n.d.-c). *Periodic trends - college chemistry*. Www.varsitytutors.com. Retrieved May 2, 2022, from https://www.varsitytutors.com/college_chemistry-help/periodic-trends

Varsity Tutors. (2012). *Balancing chemical equations - college chemistry*. Www.varsitytutors.com. https://www.varsitytutors.com/college_chemistry-help/balancing-chemical-equations

Varsity Tutors. (2015a, June 15). *Covalent bonds - MCAT physical*. Www.varsitytutors.com. https://www.varsitytutors.com/mcat_physical-help/covalent-bonds

Varsity Tutors. (2015b, July 29). *Bonding and forces - AP chemistry*. Www.varsitytutors.com. https://www.varsitytutors.com/ap_chemistry-help/compounds-and-molecules/bonding-and-forces?page=5

Varsity Tutors. (2016). *Covalent bonds and hybrid orbitals - MCAT biology.* Www.varsitytutors.com. https://www.varsitytutors.com/mcat_biology-help/covalent-bonds-and-hybrid-orbitals?page=1

Varsity Tutors High School Chemistry. (n.d.). *Help with double-replacement reactions - high school chemistry.* Www.varsitytutors.com. Retrieved May 20, 2022, from https://www.varsitytutors.com/high_school_chemistry-help/help-with-double-replacement-reactions

Vennesland, B., Stotz, E. H., & The Encyclopaedia Britannica. (2019). *Biochemistry | definition, history, examples, importance, & facts.* In Encyclopædia Britannica. https://www.britannica.com/science/biochemistry

Wade, L. G., & The Encyclopaedia Britannica. (2019). *Alcohol | definition, formula, & facts.* In Encyclopædia Britannica. https://www.britannica.com/science/alcohol